Ultratrace Metal Analysis in Biological Sciences and Environment

Terence H. Risby, EDITOR

Pennsylvania State University

A symposium sponsored by the
Division of Analytical Chemistry
at the 174th Meeting of the
American Chemical Society,
Chicago, Illinois, August
29–30, 1977.

ADVANCES IN CHEMISTRY SERIES **172**

AMERICAN CHEMICAL SOCIETY
WASHINGTON, D. C.   1979

180277   574.1921

U 47

Library of Congress CIP Data

Ultratrace metal analysis in biological sciences and
environment.

(Advances in chemistry series; 172 ISSN 0065-2393)

Includes bibliographies and index.

1. Metals—Physiological effect—Congresses. 2. Metals—Analysis—Congresses. 3. Trace elements—Analysis—Congresses.
I. Risby, Terence H. II. American Chemical Society.
Division of Analytical Chemistry. III. American Chemical Society. IV. Series.

QD1.A355 no. 172   [QP532] 540'.8s   [574.1'921]
                                                      78-31903
ISBN 0-8412-0416-0   ADCSAJ 172   1–264 1979

Advances in Chemistry Series

Robert F. Gould, *Editor*

FOREWORD

ADVANCES IN CHEMISTRY SERIES was founded in 1949 by the American Chemical Society as an outlet for symposia and collections of data in special areas of topical interest that could not be accommodated in the Society's journals. It provides a medium for symposia that would otherwise be fragmented, their papers, distributed among several journals or not published at all. Papers are reviewed critically according to ACS editorial standards and receive the careful attention and processing characteristic of ACS publications. Volumes in the ADVANCES IN CHEMISTRY SERIES maintain the integrity of the symposia on which they are based; however, verbatim reproductions of previously published papers are not accepted. Papers may include reports of research as well as reviews since symposia may embrace both types of presentation.

CONTENTS

PREFACE

Metals at various concentrations play significant roles in many biological processes and these levels are dictated by the concentration of the protein, enzyme, etc. contained in the organism. Often the concentrations of the high molecular weight biochemical moieties are low and therefore the concomitant metal concentrations will be very low. With the increasing capability for the separation and purification of enzymes, etc. and with the improving sensitivities of analytical methodologies, the knowledge of the roles of metals in biochemical processes is continually expanding. The concentration of the metal in the organism is also critical since it is possible that at one concentration the metal can be essential whereas at a higher level this same metal can cause a serious metallic imbalance in the organism. Often the range between required and excessive concentration of a metal is not well defined. In addition to the discovery of new essential metals, other metals with known requirements are found to be involved in multiple processes.

In the area of the environment, metals are considered to be significant since they are, by definition, nonbiodegradable and are retained in the ecosystem indefinitely. Metals are found in air, water, and soil as a result of both natural and anthropogenic sources. Therefore organisms can receive body burdens via various assimulation routes. The symptoms of acute metal toxicity are generally well known, although often misdiagnosed, but the symptoms of long-term exposure to low-level concentrations of metals have not been established. This latter problem currently is being studied by many researchers since these data will enable environmental epidemiological studies to be performed.

This preamble exemplifies the need for cooperation between biochemists or environmental chemists with analytical chemists. This book attempts to show areas in which the collaboration between various disciplines will be the most productive.

Department of Chemistry   TERENCE H. RISBY
The Pennsylvania State University
University Park, Pennsylvania
September 1978

Present Status and Future Development of Trace Element Analysis in Nutrition

WALTER MERTZ

U.S. Department of Agriculture, Agricultural Research Service,
Nutrition Institute, Beltsville, MD 20705

This discussion examines the recent progress of nutritional trace element research and its implications for trace element analysis. Elements recently identified as essential are present in low concentrations for which analytical methods are not yet reliable. Biological availability of trace elements depends on chemical form and on interactions with other inorganic and organic constituents of the diet. Therefore, information on elemental species is required, in addition to quantitative data. Finally, the demonstration of essential functions of trace elements previously known only for their toxicity necessitates establishing safe ranges of intake, free from danger of chronic toxicity but sufficient to meet human needs.

Trace element nutrition research is in a phase of rapid development. This is documented by the discovery of essential functions for the "new" trace elements in the past seven years (*1*) and by the increasing knowledge of marginal or pronounced deficiencies in humans of such elements as iron (*2*), zinc (*3*), chromium (*4*), and perhaps copper (*5*) and selenium (*6*). Although some of these deficiencies occur only in special age and sex groups or under unusual conditions, their existence has aroused public and scientific concern for the exact definition of human trace element needs and for the assessment of trace element nutritional status. These two major challenges for human trace element research cannot be met without proper analytical support. I will attempt to describe the present status of trace element analysis as it relates to nutrition research and to point out some new concepts that might well influence the future direction of trace element analysis.

0-8412-0416-0/79/33-172-001$05.00/0

Present Status

Trace elements fluorine, chromium, iron, copper, zinc, arsenic, selenium, molybdenum, cadmium, iodine, mercury, and lead are of public health concern because of suspected under-exposure or over-exposure (7, 8). Demands on analytical chemists for reliable analyses range from concentrations of micrograms/gram (e.g., iron) to less than 1 ng/g (e.g., chromium). Chemically defined aqueous solutions of trace elements within this range can be analyzed reliably by many different laboratories. The reliability with regard to accuracy, precision, and specificity declines drastically when foods or biological specimens (human and animal tissues) are analyzed. The steps involved in collection, preparation, and digestion of biological samples increase the probability for contamination of samples and for loss of the element under analysis by formation of refractory species or by volatilization. Each of the thousands of food and tissue samples has different matrix effects; even the urine of one individual can differ substantially in its concentration of salts from one day to another. An interlaboratory comparison on a bovine liver Standard Reference Material by the International Atomic Energy Agency revealed substantial differences in the results reported by various laboratories, ranging from 1 to 500 for the lowest and highest value for chromium to ± 30% around the certified value for zinc (9). Even the relatively small degree of imprecision for zinc is unacceptable to nutritionists. For example, the zinc content of typical daily diets in the United States apparently ranges between 10 and 12 mg (10). The Recommended Dietary Allowances, the amount considered adequate for most healthy human subjects, is 15 mg (10). In the analysis of the daily food intake, a 30% error in one direction would suggest that average intake is right at the Recommended Allowances, and the same error in the other direction would suggest that the average intake is deficient and would call for remedial measures such as zinc enrichment. Such errors also make it difficult, if not impossible, to compare data obtained by different investigators or to obtain agreement among nutritionists on which blood, urine, or tissue levels should be considered adequate or deficient. These problems do not impair the validity of comparative measurements obtained within one laboratory, but they are not compatible with large national or international studies. The only feasible solution to this problem would be the universal use of Certified Standard Reference Materials for the elements in question. Ideally, the composition of the material under analysis should be similar to that of the standard material. Ideal standards are not yet widely available, but the number and diversity of Standard Reference Materials probably will increase through the efforts of the National Bureau of Standards. At that time, scientific journals should insist on the proper use of these reference materials whenever absolute analytical trace element data are to be published. Some biological reference materials are available from the National Bureau of Standards (*11*), the International Atomic Energy Agency (*12*), and the U. K. National Physical Laboratory (Division of Chemical Standards), Teddington, England.

Food Tables

The only essential trace element for which food compositional data of some extent and reliability exist is iron (*13*). However, even assuming that the existing data are reliable, they are of limited use to nutritionists because biological availability differs greatly among iron species in foods (*14*). This will be discussed in more detail in a subsequent chapter. Food composition data for iodine are very limited. Environmental exposure of man to iodine is increasing, in part because· of the iodization of table salt but also because of the high iodine intake from adventitious sources. Concern has been expressed about potential over-exposure to this element with the concomitant risk of marginal toxicity. Systematic studies of the iodine content in modern foods are clearly needed for assessment of a potential risk (*15*).

The 1974 edition of the Recommended Dietary Allowances established an allowance for zinc of 15 mg/day for adults (*10*). Several preliminary studies of daily zinc intakes suggested that this recommended allowance is not normally met by ordinary diets unless inordinate amounts of food are consumed; yet, comprehensive composition tables for zinc do not exist. Zinc concentrations in foods differ widely among laboratories (*16*). Data are too variable to serve as a basis for important decisions now being discussed (for example, whether or not zinc fortification of certain foods should be initiated). The U.S. Recommended Daily Allowances for copper are stated as 2 mg for adult persons. Early analyses suggested that copper intakes from U.S. diets ranged from 2 to 5 mg per day, but more recent analyses of diets have detected intakes of no more than 1 mg per day (*17, 18*). Systematic tables of copper in foods are lacking. For iron, iodine, zinc, and copper, problems of analysis are not insurmountable, and our knowledge of those elements in foods should continue to progress.

In contrast, the remaining trace elements of nutritional interest are silicon, vanadium, chromium, manganese, nickel, arsenic, selenium, molybdenum, tin, and perhaps cadmium. These elements present serious problems of analysis in the concentration range that is of interest to the nutritionist. Only a few specialized laboratories have developed expertise for the analysis of these elements, considerable discrepancies still exist among laboratories, and standardized reference materials are available for only a few elements. Consequently, food composition data are few and, except for selenium and manganese, of dubious value. An appraisal of the nutritional status of the population for these elements would depend upon the development of reliable methods so that valid compositional data could be compiled. For work with those trace elements whose concentrations in biological materials are very low, "clean room" techniques may become necessary for routine work. Finally, it is likely that the demands on the analytical chemist will increase as new trace elements are added to the list of those that are now known or suspected to be essential. It is reasonable to assume that essential trace elements still to be discovered are active in very low concentrations and, therefore, could present a new set of problems to the analytical chemist.

New Concepts Influencing Trace Element Analysis

Four new concepts that have emerged from recent trace element nutrition research could influence the future development of trace element analysis.

Concept of Optimal Range. Before the discovery of their essential roles, selenium, chromium, and arsenic were known for their toxic effects. The toxicology, acute and chronic, of cadmium was thoroughly investigated; on the other hand, recent work suggests that cadmium at low concentrations might be essential to animal species (19). Even lead might have a growth stimulating effect in animals that are kept in an ultra-clean environment, according to preliminary work (1). These observations are incompatible with the concept of "toxic" and "beneficial" trace elements. Toxicity is not an attribute of any one element, but rather of an excessive concentration range. Every nutrient in sufficiently excessive concentrations can be toxic; conversely, any of the trace elements now known for their toxicity might be shown in the future to have an essential function at low concentrations (20). Low tolerances have been established to limit the accumulation of "toxic" elements in the environment to "safe" levels; for selenium and arsenic (potential carcinogens), zero tolerances are considered desirable by some. Selenium enrichment of feedstuffs for some animal species is prohibited by law, even in regions with a substantial incidence of selenium deficiency. The demonstration of essential functions for selenium and arsenic has made it necessary to replace the concept of "zero tolerance" by the more scientific idea of "safe ranges of exposure" which takes into account the fact that a deficiency of an essential element is as incompatible with health as is toxicity. The range of safe intakes is defined as meeting the requirements of individual species without causing even the slightest signs of toxicity. Physiologically, this safe range comprises intakes that fall well within the ability of homeostatic control mechanisms to maintain optimal tissue concentrations through decreased intestinal absorption and increased excretion of intakes in excess of that required. The range of safe intakes is much lower than the range of even mild chronic toxicity so application of safe intake puts substantial demands for increased sensitivity and precision on the analytical chemist.

Reliable measurements of arsenic and cadmium in a concentration range of concern to the toxicologist might be of little interest to the nutritionist. Investigation of the mechanisms of essentiality of these elements would require increases of sensitivity of several orders of magnitude over those required for the mechanisms of toxicity.

Implications of Trace Element Interactions for Analytical Chemistry. Underwood stated that one and the same copper intake of animals can result in deficiency, a normal state, or toxicity, depending on the occurrence of molybdenum in the diet. That statement might well apply to other trace elements and might be valid for all animal species and man. The interaction of copper, molybdenum, and sulfur has been studied extensively in animals; it is of substantial importance in molybdenum-rich provinces in the USSR where a gout-like syndrome in man has been reported when the copper intake is relatively low but not when it is high (21). The new knowledge of the interaction of selenium with arsenic, cadmium, and mercury is another example, not only for its toxicological but also its nutritional implications (22). Selenium protects experimental animals against cadmium poisoning. The mechanism of this protection is unknown, but it does not function through increased excretion of cadmium. This latter metal is accumulated and retained by the organism as a consequence of selenium administration to a much higher degree than in the absence of selenium. Consequently, analytical data for cadmium alone have little biological meaning, unless they are accompanied by values for selenium. These latter values determine whether or not a given cadmium concentration in a food or tissue is toxic. Although much of our present knowledge of a selenium–cadmium interaction is derived from acute studies, recent work demonstrated similar effects in long-term chronic studies with very low dietary levels of these trace elements. A similar interaction was shown for selenium and mercury. The above examples indicate that toxic effects of heavy metals depend on the concentrations of selenium present; conversely, they also indicate that the selenium requirement is determined by the amount of heavy metals in diet and environment. Many other interactions are known (7, 8), and all of these directly affect the interpretation of analytical data.

Biological Implications of Chemical Forms. The biological availability of many trace elements is influenced by their valence state. Ferrous iron is believed to be more readily available than the ferric form, and selenium is better absorbed in its high oxidation state than in its lower ones. The organism is able to oxidize or reduce some, but not all, trace elements to their biologically active form. It is important, therefore, to determine the valence state in biological material, particularly in those cases where great differences of availability or toxicity exist, as in the case of chromium or of mercury.

Even more important than valence is the chemical form in which a trace element is present. This is clearly evident when functional aspects in the organism are considered. The zinc species in albumin, in macroglobulin or low molecular weight ligands, in zinc enzymes, or in bones all represent different pools of greatly divergent biological characteristics and significance. The iron in hemoglobin in the blood has an entirely different diagnostic meaning than the iron in serum ferritin. The total cobalt concentration in serum is of little interest in monogastric species whereas a small fraction of the total that is present as vitamin B_{12} has great importance for health. One of the great challenges of nutritional trace element analysis is the identification of those chemical species of trace elements that are meaningful indicators of the nutritional status of the organism. These species have been identified for iodine, cobalt, and iron but not for the remaining essential elements. The lack of adequate indicator species for that latter group is a major obstacle to the assessment of nutritional status of individuals and population groups.

Similar demands for speciation of trace elements exist for food analysis. Substantial differences in the biological availability are known for several essential elements and depend on the form in which they are present in the diet. The chemical bases for these differences are known for cobalt, iron, and chromium but not for zinc, copper, and selenium. The importance of speciation in food analysis is best demonstrated by the example of iron. That element, when part of heme compounds, is well absorbed, and there is little influence on the absorption by other factors in the diet. Nonheme iron, on the other hand, is not readily absorbed and, in addition, is subject to many influences from dietary ingredients; those influences are poorly understood and probably not completely known (14).

Calculations of heme and nonheme iron and dietary manipulations of these two categories are believed to offer the means to improve the iron nutritional status of a large number of people in the United States and abroad. Yet, no information concerning the concentration of heme and nonheme iron in foods is available, and the calculations have to be based on the estimate of an average iron content of meats of approximately 40% heme and 60% nonheme iron.

Work on the speciation of iron and zinc in vegetable products is just beginning (23), and some progress is being made on the speciation of chromium (24) but not enough data have been accumulated to be of interest to the nutritionist. Speciation analysis has produced impressive advances in the toxicology of heavy metals. Similar advances can be expected in trace element nutrition if the difficulties of methodology can be overcome.

Influence of Food Composition on Trace Element Action. The relation between food composition and the action of trace elements is perhaps not an immediate part of trace element analysis, but it provides the background for the interpretation of analytical data, i.e., for the ultimate judgement of whether the result of an analysis represents a deficient, optimal, or toxic concentration. Even if all the criteria previously discussed have been met (the analysis is reliable, interacting elements are known, and the chemical species have been identified), the determination of the biological implication still must take into account one additional factor—the composition of the diet. The toxicity of many substances is greater in purified diets than in those containing natural materials and fiber, whereas the biological availability of zinc and iron is decreased by the fiber content. Iron, chromium, and zinc, but not selenium, are better available from animal than from vegetable products. The availability for absorption of nonheme iron is enhanced significantly by the presence of ascorbic acid in an iron-containing meal, and a similar enhancement of nonheme iron absorption is produced by the presence of meat, attributed to an unidentified "meat factor."

Although the identification and analysis of such modifying factors go beyond the scope of trace element analysis, the importance of these factors should be recognized. The complexity of the process that leads from the first step of trace element analysis to the final statement of biological implication necessitates the close collaboration between the analytical chemist and the life scientist. The chemist should not be considered the provider of data, nor the biologist the interpreter of results; rather, both scientists must be aware of the whole process—its complexity and its difficulties (25). Only through this collaboration can the enormous amount of trace element analytical data be put in order and be interpreted properly. On this whole process depends the progress of trace element nutrition research and the improvement of the nutritional status of man.

Literature Cited

1. Schwarz, K., "New Essential Trace Elements (Sn, V, F, Si): Progress Report and Outlook," *in* "Trace Element Metabolism in Animals—2," W. G. Hoekstra, J. W. Suttie, H. E. Ganther, W. Mertz, Eds., pp. 355–380, University Park, Baltimore, 1974.
2. Food and Nutrition Board, "Extent and Meanings of Iron Deficiency in the U. S.," National Academy of Science, Washington, D. C., 1971.
3. Food and Nutrition Board, "Zinc in Human Nutrition," National Academy of Science, Washington, D. C., 1970.
4. Mertz, W., "Chromium as a Dietary Essential for Man," *in* "Trace Element Metabolism in Animals—2," W. G. Hoekstra, J. W. Suttie, H. E. Ganther, W. Mertz, Eds., pp. 185–198, University Park, Baltimore, 1974.
5. Klevay, L. M., Forbush, I., "Copper Metabolism and the Epidemiology of Coronary Heart Disease," *Nutr. Rep. Int.* (1976) **14**, 221–228.
6. Levander, O., "Selenium and Chromium in Human Nutrition," *J. Am. Dietetics Assoc.* (1975) **66**, 338–344.
7. Subcommittee on the Geochemical Environment in Relation to Health and Disease, "Geochemistry and the Environment," Vol. I, National Academy of Science, Washington, D.C., 1974.
8. Ibid., Vol. II, National Academy of Science, Washington, D.C., 1977.
9. WHO/IAEA Joint Research Programme on Trace Elements in Cardiovascular Disease, *Research Coordination Meeting, Vienna, February 19–23, 1973.*
10. Committee on Dietary Allowances, "Recommended Dietary Allowances," 8th ed., National Academy of Science, Washington, D. C., 1974.
11. Cali, J. P., Mears, T. W., Michaelis, R. E., Reed, W. P., Seward, R. W., Stanley, C. L., Yolken, H. T., Ku, H. H., "The Role of Standard Reference Materials in Measurement Systems," *in* "National Bureau of Standards Monograph 148," Washington, D. C., 1975.
12. Information Sheet **LAB/243**, Int. At. Energy Agency, Vienna, 1977.
13. "Composition of Foods," Agric. Handbook No. **8**, U.S. Department of Agriculture, Washington, D. C., 1963.
14. Monsen, E. R., Hallberg, L., Layrisse, M., Hegsted, D. M., Cook, J. D., Mertz, W., Finch, C. A., "Estimation of Available Dietary Iron," *Am. J. Clin. Nutr.* (1978) **31**, 134.
15. Life Sciences Research Office, "Iodine in Foods: Chemical Methodology and Sources of Iodine in the Human Diet," Fed. Am. Soc. Exp. Biol., Maryland, 1974.
16. Schlettwein-Gsell, D., Mommsen-Straub, S., "Spurenelemente in Lebensmitteln," *Int. J. Vitam. Nutr. Res.* (1973) Supplement No. 13.
17. Wolf, W. R., Holden, J., Greene, F. E., "Daily Intake of Zinc and Copper from Self-Selected Diets," *Fed. Proc., Fed. Am. Soc. Exp. Biol.* (1977) **36**, 1175.
18. Klevay, L. M., Reek, S., Barcome, D. F., "United States Diets and the Copper Requirement," *Fed. Proc., Fed. Am. Soc. Exp. Biol.* (1977) **36**, 1175.
19. Schwarz, K., Spallholz, J., "Growth Effects of Small Cadmium Supplements in Rats Maintained under Trace-Element Controlled Conditions," *Fed. Proc., Fed. Am. Soc. Exp. Biol.* (1976) **35**, 255.
20. Mertz, W., "Defining Trace Element Deficiencies and Toxicities in Man," *in* "Molybdenum in the Environment, Vol. I," W. G. Chappell, K. Kellog Petersen, Eds., pp. 267–286, Dekker, New York, 1976.
21. Kovalskij, V. V., Jarovaya, G. A., Shmavonjan, D. M., "Changes in Purine Metabolism in Man and Animals in Various Molybdenum-Rich Biogeochemical Provinces," *Z. Obsc. Biol.* (1961) **22**, 179–191.

22. Parizek, J., Kalouskova, J., Babicky, A., Benes, J., Pavlik, L., "Interaction of Selenium with Mercury, Cadmium, and Other Toxic Metals," *in* "Trace Element Metabolism in Animals—2," W. G. Hoekstra, I. W. Suttie, H. E. Ganther, W. Mertz, Eds., pp. 119–131, University Park, Baltimore, 1974.
23. Ellis, R., Morris, E. J., "The Association of Phytate with Minerals in Wheat Bran and the Bioutilization of Zinc," *Fed. Proc., Fed. Am. Soc. Exp. Biol.* (1977) **36**, 1101.
24. Toepfer, E. W., Mertz, W., Roginski, E. E., Polansky, M. M., "Chromium in Foods in Relation to Biological Activity," *J. Agric. Food Chem.* (1973) **21**, 69–73.
25. Mertz, W., "Trace Element Nutrition in Health and Disease: Contributions and Problems of Analysis," *Clin. Chem. (Winston–Salem, N.C.)* (1975) **21**, 468–475.

RECEIVED December 5, 1977.

2

Simultaneous Multielement Analysis of Biologically Related Samples with RF–ICP

FRANK N. ABERCROMBIE,[1] M. D. SILVESTER,[2]
and ROMANA B. CRUZ

Barringer Research Limited, 304 Carlingview Drive,
Rexdale, Ontario, Canada M9W 5G2

This chapter discusses the advantages and limitations of the multielement analysis of biologically related samples using induction-coupled plasma optical emission. The sample categories covered include grains, feeds, fish, bovine liver, orchard leaves, and human kidney stones. These materials have been simultaneously analyzed for copper, nickel, vanadium, chromium, phosphorus, cobalt, lead, potassium, zinc, manganese, iron, strontium, sodium, aluminum, calcium, magnesium, silicon, boron, and beryllium, often with limited amounts of sample.

The development and the acceptance of the Induction-Coupled Argon Plasma (ICAP) as an analytical tool since the reports from Greenfield (1) and Fassel (2) have been monitored by scientists in many disciplines. Publications describing instrumentation design, performance characteristics, and analytical applications during the early years originated primarily from the Albright and Wilson and the Iowa State University researchers. More recently, review articles have appeared describing the various aspects of ICAP emission spectroscopy (3, 4, 5, 6, 7). The performance claims inferred that the ICAP would become a basic trace analysis tool for the disciplines of geological, metallurgical,

---

[1] Current address: Montana Bureau of Mines and Geology, Montana Tech., Butte, MT 59701.
[2] Current address: CSA Ltd., 55 Holmdene Close, Beckenham, Kent, U.K.

agricultural, medical, and environmental sciences. The development and acceptance of the technique was accelerated when Applied Research Laboratories (1974) and Jarrell Ash (1975) offered commercial, direct-reading emission spectrometers interfaced to ICAP sources. (Prior to this event, modular instrumentation interfacing was necessary if a laboratory intended to use an ICAP either for research or analysis.) From this current time perspective, it is possible to note that the performance attributes and analytical utility of the ICAP claimed by Greenfield and Fassel have been realized. It is now conservative to suggest that the ICAP can be regarded as an accepted analytical tool. Several commercial instrumental versions are available with more versions in various stages of planning and manufacturing. Many industry groups and most technically oriented government agencies use plasma systems within their laboratories.

The first industrial application of an induction-coupled plasma occurred within a custom analytical laboratory. At the present time, in Canada alone, there are five commercial laboratories with ICAP systems. Within our own organization, the two systems were acquired in 1974 and 1975. At that time there was a confidence that the ICAP would become a key instrument for the service facility. Since that time, extensive experience which reinforces this view has been obtained in the routine application of the ICAP to diverse analytical problems and samples.

In this chapter, the application of the ICAP to biologically related samples is discussed. Analytical data are presented for several diverse sample types and reference standards which demonstrate the suitability of one ICAP for the analysis of biologically related samples.

Two reports (8, 9) have shown potential applications for ICAP analysis in health and related areas. The data presented in the extensive report by Dahlquist (10) also supports this viewpoint. While only one specific example is presented for health-related specimens in this chapter, an implied general area of potential application is intended. Human lung, brain, heart, hair, and blood specimens have been analyzed in this laboratory but the nature of this work must remain undisclosed at this time. Other samples which have been successfully analyzed by ICAP but do not bear a detailed discussion in this report include herring gull egg shells, egg contents and whole body homogenates, rat bodies, pet food, mollusks, wheat straw and grain, tree seeds, and various legumes. The analytical results reported here, where possible, are compared with certified values or values obtained by other analytical techniques. In a few examples, however, the data are presented to demonstrate a trend; in these instances values are for information only as no reference data are available by alternate techniques.

Experimental

Digestions. After the initial sample preparation, such as surface cleaning, size reduction, homogenizing, etc. (where required), a wet-ashing acid digestion is used. To one gram of sample in a treated borosilicate beaker, 8 mL Ultrex nitric acid (J. T. Baker, Canlab, Toronto, Ontario) and 2 mL Aristar perchloric acid (British Drug Houses Chemicals, Toronto, Ontario) is added. (The borosilicate glassware is treated by soaking overnight at room temperature in 1:1 concentrated nitric–sulfuric acid.) The beaker is covered with a watch glass; the mixture is allowed to stand for one hour, and then is heated over medium heat on a hot plate. The temperature should be adjusted to maintain a steady boiling and refluxing for 1–2 hr. At the end of this period there should be a clear solution. The watch glass is removed to evaporate all excess acid. (If brownish coloration appears toward the end, indicating incomplete breakdown of the sample, more nitric acid should be added with refluxing repeated.) The solution is evaporated to dryness. The residue is then taken up in 5 mL of $0.5N$ HCl (Ultrex) and diluted to a suitable final volume, usually 10 mL. A greater dilution is used to decrease the calcium and magnesium concentration to within the dynamic range of the instrument or, for the case of some samples, to completely dissolve the potassium perchlorate present.

Instrumentation. A 32-element ICP–OES system was used throughout. A description is provided in Table I. Throughout the course of the work the spectrometer was modified as recommended by the manufacturer. The primary modification relative to this work was the installation of interference filters and photomultiplier masking assemblies. The data presented in Tables II, III, V, VI, and VIII were obtained after the interference filter and photomultiplier modification.

The main Barringer Research modification was the addition of a monochromator to provide a dedicated channel for potassium at the 766.49-nm line which is not within the wavelength region of the QA-137 and 1920 L/mn grating combination.

Spectral Lines Used (nm)

Ag	328.07	Fe	259.94	P	214.91
Al	308.21	K	766.49	Pb	220.35
B	249.77	Mg	279.55	Sr	407.77
Ca	315.89	Mn	257.61	Ti	334.95
Cd	226.50	Mo	386.41	V	292.40
Co	345.35	Na	330.23	Zn	202.55
Cu	324.75	Ni	231.60		

Calibration and Solution Analysis. Calibration is accomplished by recording the emission signals from a series of 15 solutions containing varied concentrations of the elements of interest. The combination of elements in each of the individual calibration solutions is selected to include only elements that do not cause chemical or spectral interferences with the other elements within the same solution. (For example, lead and zinc calibration solutions do not contain calcium and magnesium salts to avoid the spectral interference from calcium and magnesium.

Table I. Experimental Conditions

Generator
  output power employed   1600 W
  frequency   27.12 MHz

Gas
  argon   vapor from bulk liquid
  coolant   10 L/min
  plasma   1 L/min
  aerosol   1 L/min

Sample uptake
  rate   2–2.5 mL/min
  mode   cross-flow pneumatic, with Scott (*11*)
                                 chamber

Spectrometer   Model QA-137 (Applied Research Labora-
                                 tories, Sunland, CA)
  configuration   1-m Paschen–Runge mount
  grating   1920 rulings/mm, blazed at 350 nm
  reciprocal linear dispension 0.48–0.52 nm/mm, first order
  detectors   R300 (Hamamatsu Corp., Middlesex, NJ)
  entrance slit   12 μm
  exit slits   50 μm
  plasma observation height   16 mm above load coil, 4-mm vertical
                                 section

Monochromator
  optics   50-mm focal length, 25.4 mm quart, 1:1
                                 image focused on a 500-mm fiber
                                 bundle, placed immediately in front of
                                 the entrance slit
  monochromator   ½ M Ebert (Jarrel–Ash Division, Fisher
                                 Scientific Co., Boston, MA)
  slits   50 μm
  grating   ruling information unknown, near-infrared
                                 blaze
  photomultiplier   R787 (Hamamatsu Corp., Middlesex, NJ)

Lead and silver calibration solutions do not contain other metals as the chloride salts. Also, sodium and potassium calibration solutions do not contain molybdenum, phosphorus, silicon, and chromium, which are used as the potassium and sodium salts). The acid concentration of the samples is adjusted to within ±10% of the concentration of the calibration solutions. This procedure effectively minimizes solution uptake rate and nebulization efficiency effects which have been reported (*10, 12, 13*).

The emission spectrometer is interfaced to a programmable calculator (Model 9830, Thermal Printer 9866, Hewlett–Packard, Mississanga, Ontario). During calibration raw millivolt data of the matrix acid blank are subtracted from the standard data. Slopes and intercepts are obtained through a second-order polynomial regression. During analysis all data listed on the thermal printer and stored on the ⅛-in. magnetic tape cassettes is in solution concentration units with appropriate sample identification and dilution information. The data is relisted, reporting the sample concentration for the elements of interest in an appropriate tabular format.

Results and Discussion

Current commercial ICAP systems are designed for relatively non-viscous liquid samples and standards. Thus, solid samples must be dissolved or leached via chemical procedures. The procedures which are commonly used include: digestion–oxidation in a fluorocarbon polymer bomb containing an oxidizing acid; low-temperature plasma ashing followed by acid leaching of the residue; controlled temperature dry ashing in a muffle furnace (approximately 450°C) followed by acid leaching of the residue; and a wet digestion–oxidation by a strong oxidizing acid reflux. The digestion used throughout this chapter is a controlled reflux in $HNO_3/HClO_4$. (Ward and Marciello (14) have recently compared several digestion procedures and $HF/HNO_3/HCl$ reflux seems to be a superior digestion for the analysis of samples containing silicon and elements which tend to form silicates.) While the $HNO_3/HClO_4$ digestion had been successfully used in the past at Barringer for several elements by classical atomic absorption sequential analysis, no work had demonstrated the applicability of the digestion or the capability of the staff to carry out simultaneous trace multielement analyses on biological samples. The average number of elements determined per sample increased from five for atomic absorption spectroscopy to approximately 20–25 for the ICAP approach.

Standard reference materials were the first samples analyzed to evaluate the preparation chemistry and instrumental capability. Evaluation of the initial analytical results demonstrated that a greater degree of analytical expertise was required to ensure precise and accurate multielement analysis. In many instances, lack of agreement with the certified value was traced to a chemical incompatibility in the digestion procedure

Table II.   Comparison of Barringer (BRL)

	Ni (ppm)	P (%)	Na (ppm)
Orchard leaves			
NBS 1571	1.3 ± 0.2	0.21 ± 0.01	82 ± 6
BRL	1.8 ± 0.2	0.20 ± 0.05	140 ± 12
Bovine liver			
NBS 1577	—	—	2430 ± 130
BRL	< 0.5	1.16 ± 0.03	2320 ± 300

[a] Barringer Research samples were digested in $HNO_3/HClO_4$; NBS results reported are for $HNO_3/HClO_4$.

or to contamination. Water and acids of the highest purity were required to avoid high reagent blanks relative to the trace element content of most organic samples. The results presented in Tables II and III demonstrate the quality of data which can be obtained from the simultaneous multielement analysis of NBS orchard leaves and bovine liver using one digestion for all elements. In general, satisfactory agreement is obtained for all elements presented in the tables with only a few elements bearing special mention. The high sodium result for orchard leaves is probably attributable to a spectral interference and the limited detection limit of the 330.23-nm line used.

It is worth mentioning that certified reference data were not available for the elements aluminum, titanium, vanadium, boron, nickel, and phosphorus at the time of analysis. This is a frequent problem in multielement analyses. It is difficult to obtain standard biological reference materials similar to the samples being analyzed with more than but a few certified analytical results. The lack of standard reference material limits quality assurance review of the data and diagnosis of analytical problems. The recent issue of new standards such as spinach leaves and brewers yeast will improve the material variety. Certification of more elements in the existing standards will also greatly assist the analyst.

The lack of reference data also is apparent in Table IV, which compares ICAP analysis results with the values supplied by the International Atomic Energy Agency (IAEA) for a number of reference materials. In most instances where an IAEA value is given, the ICAP result generally is in close argeement. Only a few unexplained discrepancies occur. The precision of the ICAP values is worse than what is normally expected; however, it is similar in magnitude to the precision of the data used to assign the reference values. This precision may indicate a problem with sample homogeneity.

For the calcined bone analysis, review of the data for 20 elements for each replicate sample presented patterns of elements which tracked

ICP OES Results with NBS Values[a]

Sr (ppm)	Zn (ppm)	Al (ppm)	Ti (ppm)	V (ppm)
(37) 35 ± 3	25 ± 3 21 ± 1	— 146 ± 20	— 6.6 ± 0.5	— 2.2 ± 0.1
(0.14) 0.30 ± 0.06	130 ± 10 143 ± 19	— 6 ± 2	— 1.7 ± 0.2	— 0.6 ± 0.1

Table III. Comparison of Barringer (BRL)

	Ca (%)	Cu (ppm)	Fe (ppm)
Orchard leaves			
NBS 1571	2.09 ± 0.03	12 ± 1	(270)
BRL	2.08 ± 0.06	12.2 ± 1.1	245 ± 35
Bovine liver			
NBS 1577	(123)	193 ± 10	270 ± 20
BRL	134 ± 18	190 ± 10	262 ± 13

[a] Barringer Research samples were digested in $HNO_3/HClO_4$; NBS results reported are for $HNO_3/HClO_4$.

one another. In multielement analysis, these patterns may often reveal sample or procedural problems such as heterogeneity, extraction efficiency, or transfer errors. The information sheet supplied for A-3/1 noted that the chromium content (IAEA 683 ± 241 ppm vs. BRL 673 ± 391 ppm) was a suspected contamination during grinding. (The different precision values may result from different analytical sample masses.) The individual percent relative standard deviations of four replicates for the elements were: Al 19, Fe *36*, Ca 20, Mg 25, Si *56*, Na 28, K 24, Ni *43*, Cu 28, Zn 26, Cr *61*, Co 28, Cd *34*, Ag 24, Sr 23, V 24, Be 19, P 20, Ti 29, Mn *37*. In this example, the groupings of the relative standard deviations were consistent with an external contamination (alloy elements) superimposed upon a general sample heterogeneity or sample preparation problem.

Table IV. Comparison of Barringer (BRL)

	Fe (ppm)	Ca (ppm)	Mg (ppm)
Potato			
IAEA V-4	18.6 ± 4.7	—	—
BRL	19 ± 6	410 ± 110	700 ± 130
Wheat flour			
IAEA V-2/1	84 ± 17	—	—
BRL	80 ± 50	670 ± 80	1660 ± 24
Fish solubles			
IAEA A-6	565 ± 93	—	—
BRL	570 ± 80	1600 ± 120	1100 ± 90
Calcined bone			
IAEA A-3	1520 ± 360	—	—
BRL	1600 ± 560	39% ± 7%	8100 ± 2000

[a] International Atomic Energy Agency, Kartner Ring 11, P.O. Box 590, A-1011 Vienna, Austria.

ICP OES Results with NBS Values[a]

B (ppm)	Pb (ppm)	Mg (%)	Mn (ppm)
33 ± 3	45 ± 3	0.62 ± 0.02	91 ± 4
32 ± 4	47 ± 4	0.55 ± 0.03	90 ± 3
—	0.34 ± 0.08	(605)	10.3 ± 1.0
4 ± 1	< 0.5	590 ± 40	10.0 ± 1.3

The analysis of nitric–perchloric acid digests of feed samples for a local industry presented an early test of the ICAP analysis of organic samples. The results obtained for Association of American Feed Control Officials (AAFCO) feed check samples (included as quality assurance standards within the sample suite) are given in Tables V and VI. The ICAP results for iron, copper, zinc, manganese, cobalt, and potassium are all within the uncertainty limits of the certified values. While not within the uncertainty limits, the results for calcium, magnesium, sodium, and phosphorus compare with acceptable agreement for the intended application.

The precision and accuracy obtained with the ICAP analysis imply that the analysis using ICAP data at percentage levels should be a feasible alternative to the classical methods for some applications. The data

ICP OES Results with IAEA Values[a]

P (ppm)	Cu (ppm)	Zn (ppm)	Mn (ppm)	Sr (ppm)
—	4.2 ± 1.1	11.9 ± 1.3	2.4 ± 0.5	—
1600 ± 300	4.1 ± 0.2	12.5 ± 4	2.2 ± 0.5	2.1 ± 0.3
—	5.8 ± 1.1	33 ± 6	35 ± 6	—
4120 ± 250	5.6 ± 1.3	37 ± 7	35 ± 6.3	2.0 ± 0.5
—	5.25 ± 1.2	18.9 ± 2.9	4.73 ± 1.01	—
1.1% ± .2%	5.2 ± 0.5	23.7 ± 2.4	5.0 ± 0.4	4.4 ± 0.5
—	6.8 ± 2.3	183 ± 12	32 ± 5	—
16% ± 3%	8.4 ± 2.4	113 ± 30	20 ± 7	200 ± 45

Table V.   Comparative Data on

	Fe (ppm)	Cu (ppm)
7625—Certified	447 ± 42	15 ± 6
—BRL	495	12
7530—Certified	395 ± 140	15 ± 7
—BRL	310	14

[a] Certificate values and sample solutions courtesy of C. H. Perrin of Canada Packers, Research Centre, Toronto, Ontario.

Table VI.   Comparative Data on

	Ca (%)	Mg (%)
7625—Certified	3.23 ± 0.18	0.131 ± 0.094
—BRL	3.62	0.165
7530—Certified	1.04 ± 0.05	0.156 ± 0.004
—BRL	1.04	0.163

[a] Certificate values and sample solutions courtesy of C. H. Perrin of Canada Packers, Research Centre, Toronto, Ontario.

presented in Table VII demonstrates the excellent agreement obtained between ICAP and the Association of Official Analytical Chemists' (AOAC) wet chemical methods for calcium and phosphorus in feeds. Only the phosphorus result for sample C was unacceptably divergent. No attempt was made to characterize this discrepancy.

Another comparative analytical study on actual feed industry samples (nonreference material) is presented in Table VIII. These results demonstrate the dynamic range of ICAP analyses. The agreement of the ICAP results with atomic absorption and wet chemistry in the respective ranges is satisfactory for general control purposes. The data also illustrate

Table VII.   Comparison[a] of Barringer (BRL) Results ICP OES with Wet Chemistry Data on Animal Feeds

	Ca (%)	P (%)
A—Wet chemistry	12.5	11.8
—BRL	12.9	11.7
B—Wet chemistry	24.5	10.6
—BRL	23.6	10.7
C—Wet chemistry	15.3	25.8
—BRL	14.9	21.1

[a] Reproduced with permission of C. H. Perrin of Canada Packers Research Centre, Toronto, Ontario.

AAFCO Feed Check Samples[a]

Zn (ppm)	Mn (ppm)	Co (ppm)
90 ± 19	99 ± 8	5 ± 3
77	95	3.5
123 ± 12	88 ± 8	1.5 ± 0.5
124	92	1.8

AAFCO Feed Check Samples[a]

Na (%)	K (%)	P (%)
0.55 ± 0.05	—	0.687 ± 0.025
0.53	1.22	0.72
0.255 ± 0.007	1.15 ± .01	0.81 ± 0.07
0.23	1.14	0.88

the relative freedom from interferences exhibited by ICAP analyses. The atomic absorption (AA) results were seriously depressed by the presence of high levels of phosphorus. This discrepancy was easily reconciled in the data review of the multielement report. This type of reconciliation for the analyst or customer, while not routine, frequently occurs and is an important and latent benefit of multielement analysis. It is conceivable that the erroneous atomic absorption results for calcium would have been reported if other knowledge about the sample were unavailable.

Table VIII. Comparison of Wet Chemical and Atomic Absorption Analysis with ICAP

	Ca (%)			P (%)
	Wet Chemistry	BRL ICAP	BRL AA	ICAP
Wheat	—	0.03	0.034	
Corn	—	0.0065	0.006	
Barley	—	0.052	0.056	
Soya	—	0.20	0.2	
Wheat shorts	—	0.11	0.13	
Dairy ration	—	0.78	1.02	
Dog food	1.5	1.6	1.8	
Meat meal	8.0	7.1	6.6	3.2
Fish meal	9.0	8.1	6.2	3.8
Premix	23	21	14.0	9.6

Table IX.   Typical Multielement Analysis[a]

	Na (%)	K (%)	P (%)	Mg (ppm)
Nutrient broth	3.6	3.0	1.1	235
Tryptone broth	2.9	< 0.1	0.8	315
BRL nutrient broth	3.0	2.8	1.0	304

[a] Reproduced with permission of Thomas Jack, Scarborough College, University of Toronto.

The precision and accuracy of multielement ICAP analyses are ideal for nonspecific screening of material where it is necessary to determine group differences. The ICAP screening capability is often used advantageously for analytical problems where the analytical requirement is the determination of a causative agent. Multielement data will reveal whether or not basic elemental chemical differences exist between sample groupings or batches; the differences may then explain performance, quality, or elegance differences between batches. (Even the event where no elemental chemical differences exist between batches or groups is significant in that the determined elements can be eliminated as a source of a problem.) The multielement screening approach has been used successfully to characterize the source of material differences in samples of pharmaceutical raw material and products. The data presented in Table IX demonstrate the type of elemental differences obtained for bacterial culture media that exhibited dramatically different growth characteristics for identical bacterial cultures. In this instance, the ICAP advantangeously monitored essential elements as well as potentially toxic elements contained in the culture media. With these data, it was possible

Table X.   Typical Multielement Analysis[a]

	Fe (ppm)	Ca (ppm)	Mg (ppm)
Feathers			
A	11.6	100	15
B	16.5	120	16
C	8.5	80	12.5
D	8.8	100	14
Quills			
A	13.7	28.5	12.1
B	6.6	22.1	11.3
C	3.4	19.8	8.8
D	2.5	19.8	12.3

[a] Reproduced with permission from Dr. D. Carlisle, Canadian Wildlife Services, Environment Canada.

by ICP OES Bacterial Culture Media[b]

Mn (ppm)	Ni (ppm)	Al (ppm)	Cu (ppm)	Sr (ppm)	Ca (ppm)
0.43	3	3	0.7	0.11	67
1.02	3	7	1.2	3.21	602
0.19	1	< 1	1.3	0.05	80

[b] Sample dissolved in demineralized distilled water only.

to interpret the growth rates as being attributable to the trace-element content more than to the biological differences of the two growth media.

Environmental problems often benefit from the application of screening analysis. Data obtained from the analysis of whooping crane feathers are presented in Table X. (The attention to whooping cranes results from a decline in the flock population over the past few years.) In this instance, the ICAP was able to provide screening data for limited amounts of sample, as only one wing feather per bird was available. Common analytical techniques would not have been able to provide such a broad spectrum of elemental analyses as cost effectively as the ICAP approach.

(The data presented in Tables IX and X are representative of the elements for which there was a definite indication of between-sample differences or a definite information requirement requested by the end user of the analytical data. As is the norm, all analytical channels were monitored.)

Kidney stones represent another sample class ideally suited for ICAP screening analysis. The objective is to identify trace-element patterns

by ICP OES Whooping Crane Feathers

V (ppm)	Ti (ppm)	Ni (ppm)	Sr (ppm)	Cu (ppm)
.035	0.335	0.808	0.114	0.706
.042	0.428	1.05	0.093	0.661
.024	0.273	0.579	0.078	0.592
.026	0.250	0.725	0.072	0.664
.020	0.121	0.947	0.028	1.08
< .001	0.090	0.726	0.014	0.865
.002	0.091	0.583	0.013	0.878
.002	0.086	0.586	0.013	0.875

that correlate kidney stone occurrence with diet, environment, or geographic location. The small, often limited, sample mass presents a restriction for destructive single element or sequential elemental instrumentation. The kidney stone samples readily dissolve in the nitric–perchloric acid digestion. (Kidney stones often occur as oxalate salts which are readily destroyed by oxidizing acid digestion.) Typical analytical results for representative elements in kidney stones are presented in Table XI. This sample set–element combination was chosen to illustrate the type of variations which can occur in a screening study.

The commonly considered advantages of ICAP analyses are: precision, accuracy, low detection limits, and relative freedom from chemical interferences. The multielement analytical capability also implies a cost effective approach; 200 samples can easily be analyzed in an eight-hour shift. The wide dynamic range of the ICAP also reduces the work effort by minimizing multiple sample dilution requirements. The linear dynamic range was 1000 μg/mL for all elements herein except calcuim and magnesium. (Magnesium analytical curves exhibit nonlinearity for concentrations greater than 100 μg/mL for 279.55, 280.2, and 285.21 nm. The calcium analytical curves for 393.37 and 315.89 nm become nonlinear at concentrations greater than 20 and 1000 μg/mL, respectively.) The expendable operating costs (i.e., primarily argon and torches) exclusive of reagents are less than 10¢ per sample.

The requirement of only one set of ICAP operating conditions is a definite advantage. Relative to atomic absorption there is no requirement for different combinations of releasing agents, ionization buffers, varied flame conditions, and correct burner for the individual elements. This also reduces the opportunity to use incorrect parameters during analysis. An ICAP operator does not require the same level of training and supervision to ensure conformance to such a wide range of requirements and operator attention as atomic absorption. After two days of training and four days of close supervision, a fourth-year high school (mathematics

Table XI. Typical Multielement Analysis[a]

	Fe (ppm)	Al (ppm)	Zn (ppm)
#25	14.4	134	43
#27	41.6	335	358
#31	8.3	37.4	158
#38	13.1	68.6	153
#41	11.8	79	21
#51	36.7	31.4	14.8
#65	34	138	25

[a] Reproduced with permission from A. Levinson, University of Calgary.
[b] N.D. = not detected.

college preparatory) student successfully worked a second shift in the summer season with no more supervision than required for the usual operator, a B.Sc. Chemist. The supervisor of an ICAP, however, must be a competent analytical chemist who is well versed in sample preparation, data review, and quality assurance procedures.

The simultaneous multielement capability and low detection limits of an ICAP facilitates the analysis of mass-limited samples. Kniseley et al. (9) discussed this advantage with respect to small sample volumes of blood and serum. This feature was indispensable in the analysis of all specimens of human origin. Other areas where limited sample masses have been analyzed relate to materials testing, air particulates, and corrosion testing. Sample masses as low as a few milligrams have been successfully analyzed for 200-μL volumes.

For environmental, biological, or health-related investigations, the primary limitation of ICAP analysis is related to the totality of the analysis. The concentrations of specific element forms are often more important than total element content in these disciplines. ICAP analysis by direct pneumatic nebulization to date must be considered to provide only the total content of each element analyzed, independent of the forms. If a specific element form is to be determined, appropriate separation techniques must be used prior to analysis by conventional sample introduction to an ICAP.

The complexity of the ICAP emission spectra present a definite limitation to the design of an ICAP system and to the practical application to analyses (15). While this situation may be common to most or all emission spectroscopic techniques, it must be recognized in the design of ICAP instrumentation, line selection, and data review. Fortunately most spectral problems can be eliminated for most elements in most samples with the appropriate use of interference filters and computer correction techniques (10).

by ICP OES Kidney Stones

Mn (ppm)	Cd (ppm)	Ag (ppm)	Ti (ppm)	Mo (ppm)
N.D.[b]	2.5	N.D.[b]	0.71	26
0.12	1.6	N.D.[b]	1.76	55
0.51	0.5	N.D.[b]	1.27	62
0.15	0.23	0.116	1.25	61.3
N.D.[b]	1.9	N.D.[b]	0.62	28
0.24	0.5	0.50	1.00	53
N.D.[b]	N.D.[b]	N.D.[b]	N.D.[b]	N.D.[b]

Other attributes exist which are advantages or limitations, depending upon the perspective of the view. Additional attributes of ICAP analysis exist which are only requirements. Some of these are mentioned because they may not be immediately obvious. Awareness of their existence can be useful when considering the ICAP as a potential method of analysis.

The simultaneous multielement approach is often presented as advantageous and indeed it generally is. A latent benefit which has resulted is several client problems have been resolved because the report form includes the results for elements not requested by the client. The "extra" elements were related to the problem while the requested data was not relevant to the problem. Since there had been no previous reason to even be aware of the presence of the extra elements in the samples, these elements were not requested. (It is less time consuming to report all results as opposed to deleting different combinations for each client.) In a single-element approach the samples would not have been analyzed for these additional elements because of the additional cost.

The simultaneous multielement approach by ICAP will generally require the laboratory and the user to appreciate data management. An ICAP analyzing 200 samples per day for even a 25-element array will be producing 5000 individual element concentrations for review and interpretation. If any given study includes many samples, computer techniques are a definite asset. However, if the computer is to be used to the greatest extent, the data input/output must be rapid. Paper tape or magnetic tape cassette storage is not recommended. In addition, although an 8000 word memory can be used with extensive segmenting, a 16,000 word memory is recommended.

Presentation of the sample in a liquid form is a limitation with respect to chemical solubility, the labor involvement, and potential contamination. However, most analytical techniques require a liquid sample. The solution obtained from the totality of a solid sample is a homogeneous sample irrespective of the homogeneity of the original sample. The preparation of sample solutions for multielement analysis also requires complete dissolution of all elements by one procedure (if it is to be truly multielement) with minimal or no contamination. If more elements are included in the analytical matrix, it is more difficult to find a digestion procedure which will quantitatively dissolve all the elements to be determined. Freedom from sample contamination (and standard contamination) also becomes more difficult to maintain when the number of analyte elements is large. In the single-element mode, a minor chromium contamination in zinc analysis will not be a problem. In fact it will probably not even be noticed. This is not the case in multielement analyses. Contamination problems are recognized more frequently in ICAP analysis because of the large number of elements which are determined and because the ICAP is capable of trace level determinations. The requirement for contamination-free, quantitative sample digestion techniques and procedures is of utmost importance in the successful use of an ICAP system. While it is the responsibility of the instrument manufacturer to provide a quality, functionally designed instrument capable of performing in the intended application, it still remains a responsibility of the analytical chemist to decide if the instrumentation is appropriate for the application and to ensure that the laboratory functions present the appropriate analytical sample for analysis by the instrument.

Conclusions

An ICAP emission spectrometer in a commercial analytical laboratory can successfully provide accurate, precise multielement data (at major, minor and trace levels) for biological and human-related samples for many of the elements of interest for the related disciplines. The relative freedom from interferences is a very positive attribute. The analytical cost of operation is attractive whenever more than four elements must be analyzed in a sample. The inability of the experimental approach used here to provide analytical data for individual species of the elements is a definite disadvantage when this information is required. The primary requirement for ICAP-simultaneous multielement analysis is exceptionally careful analytical sample preparation methods and laboratory techniques.

Acknowledgment

We thank A. J. Loveless for his helpful suggestions.

Literature Cited

1. Greenfield, S., Jones, I. U., Berry, C. T., *Analyst (London)* (1964) **89**, 713.
2. Fassel, V. A., Wendt, R. H., *Anal. Chem.* (1965) **37**, 920.
3. Kirkbright, G. F., *Analyst (London)* (1971) **96**, 609.
4. Fassel, V. A., Kniseley, R. N., *Anal. Chem.* (1974) **46**, 1110A, 1155A.
5. Boumans, P. W. J. M., *Philips Tech. Rev.* (1974) **34**, 305.
6. Winefordner, J. D., Fitzgerald, J. J., Omenetto, N., *Appl. Spectrosc.* (1975) **29**, 369.
7. Barnes, A. M., *Anal. Chem. Fund. Rev.* (1976) **48**, 106R.
8. Greenfield, S., Smith, P. B., *Anal. Chim. Acta* (1972) **59**, 341.
9. Kniseley, R. N., Fassel, F. A., Butler, C. C., *Clin. Chem. (Winston–Salem, NC)* (1973) **19**, 807.
10. Dahlquist, R. L., Knoll, J. W., *Appl. Spectrosc.* (1978) **32**, 1.
11. Scott, R. H., Fassel, V. A., Kniseley, R. N., Nixon, D. E., *Anal. Chem.* (1974) **46**, 75.
12. Sobel, H. R., Kniseley, R. N., Fassel, F. A., Paper 339 presented at the 26th Pittsburgh Conference on Analytical Chemistry and Applied Spectroscopy, Cleveland, 1975.

13. Koirtyohann, S. R., Lichte, F. R., Presented at the 5th International Conference of Atomic Spectroscopy, Melbourne, Australia, 1975.
14. Ward, A. F., Marciello, L., Paper presented at the 29th Pittsburgh Conference on Analytical Chemistry and Applied Spectroscopy, Cleveland, 1978.
15. Larsen, G. F., Fassel, V. A., Winge, R. K., Kniseley, R. N., *Anal. Spectrosc.* (1976) **29**, 384.

RECEIVED June 8, 1978.

Health Implications of Trace Metals in the Environment

KENNETH BRIDBORD and HARVEY P. STEIN

National Institute for Occupational Safety and Health, Center for Disease Control, Public Health Service, U.S. Department of Health, Education, and Welfare, Rockville, MD 20857

The objective of this chapter is to put into perspective some of the current knowledge with respect to trace metals and their health implications. Potential adverse health effects of occupational exposures to trace metals are discussed: cancer (arsenic, beryllium, chromium, nickel, and perhaps cadmium); chronic lung disease (beryllium and cadmium); neurologic and reproductive disorders (lead and mercury); and kidney disorders (lead and cadmium). Also discussed are the National Institute for Occupational Safety and Health (NIOSH) recommended standards for occupational exposure to several trace metals, the difficulty of establishing "safe" levels of exposure (particularly for carcinogens), and problems involved in identifying toxic components of trade name products. Special attention is given to the role of chemists to help protect the public health.

The presence of trace metals in the environment, both in the workplace and in the community, has been the subject of considerable public health concern in recent years (1–15). Human exposure to trace metals occurs primarily through inhalation of air and ingestion of food and water. Concentrations of trace metals in community air and in the diet vary considerably depending upon a number of factors, including proximity to sources of trace metal emissions. The trace metals of greatest concern for the general population are those which are ubiquitous in the environment. Lead is a good example, being present in substantial quantities in the ambient air and in the diet. In general, workers comprise the

0-8412-0416-0/79/33-172-027$05.00/0

group most highly exposed to trace metals, and occupational exposure varies considerably in magnitude depending on the job situation. Estimates of the number of workers exposed to lead and to arsenic exceed one million each (7, 11). The highest exposures tend to concentrate in a defined number of industries. For example, the highest exposures to lead occur in primary and secondary smelting and in lead storage battery manufacturing and involve approximately 25,000 workers.

Health implications of low-level chronic exposure to trace metals are not clearly understood. However, the trend of scientific investigation is generally shifting from emphasis on acute toxic effects to a greater concern for the effects of long-term exposure. Although trace metals are essential nutrients, excessive exposure to trace metals has been associated with a variety of adverse effects including cancer (arsenic, beryllium, chromium, nickel, and perhaps cadmium), chronic lung disease (beryllium and cadmium), neurologic and reproductive disorders (lead and mercury), and anemia and kidney disorders (lead and cadmium) (7–15).

The largest body of information concerning the health effects of trace metals has come from studies of exposed workers. Although frequently involving much higher exposure levels than encountered by the general population, adverse effects observed among workers do provide insight into effects which can occur among the general population. Data on the health effects of arsenic, beryllium, cadmium, chromium, lead, mercury, and nickel have been reviewed and summarized in criteria documents prepared by the National Institute for Occupational Safety and Health (NIOSH) (7–13). Additional knowledge regarding these health effects also has come from community studies as well as from toxicologic studies involving experimental animals. In the case of lead, considerable knowledge has been derived from studies of children exposed to lead-based paint and to other lead sources.

Safe Levels of Exposure

As new knowledge is obtained, the perspective as to safe levels of exposure to trace metals has resulted in increasingly more stringent exposure standards both in the workplace and in the general population. The Consumer Product Safety Commission has recently reduced the maximum concentration of lead permitted in consumer paints from 0.5 to 0.06%. The Environmental Protection Agency (EPA) is currently in the process of establishing an ambient air quality standard for lead. Mercury and beryllium both have been declared "hazardous" pollutants by EPA and are regulated as such under "The Clean Air Act." Perhaps the clearest example of how "new knowledge" has resulted in reducing permissible exposure limits comes from studies of workers. Within the past two years, NIOSH has recommended the reduction of occupational exposure to a number of trace metals including arsenic, beryllium, cadmium, chromium, lead, and nickel.

An employee exposure ceiling of 2 μg arsenic per m^3 air (as determined by a 15 min sampling period) was recommended by NIOSH to replace the existing limit of 500 μg/m^3, largely because of the carcinogenic effect of arsenic (7). NIOSH recommended an exposure limit of 1 μg/m^3 for carcinogenic hexavalent chromium; the previous general recommendation for chromium had been 100 μg/m^3 (10). NIOSH has recommended that allowable exposure to nickel be reduced from a 1000 μg/m^3, 8 hr time-weighted average (TWA) to a level of 15 μg/m^3 because of nickel's carcinogenic activity, including increased risk of lung, nasal, and kidney cancer among exposed workers (13). The recommendation that maximum occupational exposure to beryllium be reduced to 0.5 μg/m^3 was based on NIOSH's assessment of the carcinogenicity of beryllium in humans (14). In these cases, for trace metals believed to be human carcinogens, NIOSH has recommended that occupational exposure be reduced to the lowest levels which can be quantitatively measured by specified sampling and analytical procedures routinely applicable to employee exposure monitoring.

A reduction in the maximum permissible lead exposure from 200 μg/m^3, the Occupational Safety and Health Administration (OSHA) standard, to a TWA of 100 μg/m^3 was recommended by NIOSH largely because of evidence showing adverse neurologic, kidney, hematologic, and reproductive effects among workers (11, 15). NIOSH also recommended that the standard for occupational exposure to cadcium be reduced from 100 μg/m^3 to 40 μg/m^3 TWA because of the effect of cadmium upon kidney function (9). At issue in this recommendation was whether or not exposure to cadmium poses a carcinogenic risk to either the lungs or the prostrate. While NIOSH did not consider the available evidence sufficient to recommend the regulation of cadmium as a carcinogen, future studies might well confirm the carcinogenic potential of cadmium. If so, a major determinant of any future cadmium standard may be the capability of sampling and analytical techniques routinely applicable to employee exposure monitoring.

Role for Chemists

Chemists play an important role in protecting the health of workers at risk of exposure to trace metals. Recommendations of exposure limits for a number of carcinogenic metals (i.e., arsenic, beryllium, chromium, and nickel) have been based upon the limitations of sampling and analytical technology routinely applicable to employee exposure monitoring rather than upon specific health effects studies. This has occurred because "safe" levels of exposure to carcinogens cannot be defined currently. Although a substance can be shown to be carcinogenic by appropriate epidemiologic and/or toxicologic data, the available information has not permitted the calculation of "safe" exposure levels. Under these circumstances, to minimize occupational exposure to carcinogens, NIOSH has generally recommended that occupational exposure to human carcinogens be no greater than the lowest levels that can be reliably measured by specified sampling and analytical procedures routinely applicable to the workplace. (In many instances, laboratory techniques do exist that can measure exposure below the recommended standard, yet these techniques were not believed to be readily available for use as routine monitoring procedures in the workplace).

If new practical analytical methods can be developed or if existing methods can be adapted for routine employee exposure monitoring, then it might be possible to recommend occupational standards with greater margins of safety for protecting health. NIOSH therefore has participated in the development of promising new techniques, including graphite furnace (atomic absorption spectrophotometry) and inductively coupled plasma (optical emission spectroscopy), in its own research laboratories as well as through external funding.

The need for improved sampling and analytical techniques for monitoring employee exposure extends to substances other than the trace metals. The "no detectable limit" philosophy for occupational exposure to carcinogens dates back to the 1974 situation with vinyl chloride in which NIOSH recommended that airborne concentrations be reduced "to levels not detectable by the recommended method" (1 ppm) (*16*). Very low maximum permissible exposure levels are likely to be recommended in the future for substances which are determined to be potential human carcinogens.

The issue of analytical techniques for exposure monitoring extends beyond the workplace. Stationary sampling equipment used for community air monitoring generally lends to the development of systems that have lower limits of detectability than those based on the personal samplers used in the workplace. This has facilitated the recommendation of standards for community exposures which generally specify lower levels than those permitted in the workplace. However, this difference reflects not only sampling and analytical technology, but also the fact that community exposure occurs 24 hours a day, seven days a week, as opposed to occupational exposure which is usually limited to 40 hours per week. Other factors, including the existence of more high-risk groups in the general population (e.g., the very old and the very young), also have tended to reduce the maximum levels of exposure permitted in the community as compared with the workplace. As the recommended air standards for the workplace become more stringent, however, there might be demands to further reduce community exposures to a point at which the limitations of sampling and analytical technology become factors determining the levels at which standards are ultimately set. Arsenic and beryllium are two current examples where this situation might occur.

The availability of adequate control technology to reduce emissions of hazardous materials into the workplace and into the general environment also has been a severe limiting factor in reducing exposure. It is easier to design controls into the basic process than to retrofit controls once a chemical plant has been built. Arsenic and lead are examples of trace metals where the lack of adequate control technology severely hampers the ability to reduce both occupational and community exposure. It is unfortunate that both chemists and chemical engineers frequently do not receive formal training in the health hazards of chemical processes and in various control techniques. Chemists, working together with engineers and health scientists, could then better contribute to the development of control technology and to the modification of basic processes, thereby helping to contain fugitive emissions.

Chemists also can play an important role in helping resolve the trade name product issue. It is extremely difficult for workers to protect themselves unless they know what materials they are being exposed to. A particular problem has been the existence of trade name products that do not readily reveal the presence of toxic substances. The NIOSH National Occupational Hazard Survey, in a two-year field survey of approximately 5,000 workplaces, identified 86,000 trade name products and has been successful to date in obtaining the formulations of approximately 50,000 of these products. Within this latter group, more than 20,000 trade name products contained federally regulated toxic substances and more than 400 contained known cancer-causing agents. The situation is complicated by trade secret designations claimed in roughly one-third of all responses to the NIOSH survey, which further obscures the presence of toxic substances in commercial products. This poses a special challenge to chemists in helping identify toxic substances in commercial products so that proper precautions can be taken to protect the worker and the general population.

Analytical chemistry also plays a crucial role in toxicologic and epidemiologic studies because of the importance of ascertaining the actual exposures of the subjects under investigation. We are all certain to benefit from the continuing contributions of the chemist and chemical engineer in elucidating the health effects of trace metals and in controlling their adverse impact.

Literature Cited

1. Horvath, D. J., "Trace Elements and Health," *in* "Trace Substances and Health—A Handbook," P. M. Newberne, Ed., Part I, Marcel Dekker, New York, 1976.
2. "Metallic Elements in Human Health," D. H. K. Lee, Ed., Academic, New York, 1972.
3. Schroeder, H. A., Darrow, D. K., "Relation of Trace Metals to Human Health," *Environ. Affairs* (1972) 2, 222.
4. Schroeder, H. A., Darrow, D. K., "Relation of Trace Metals to Human Health Effects," *Prog. Anal. Chem.* (1973) 5, 81.
5. Underwood, E. J., "Trace Elements in Human and Animal Nutrition," Academic, New York, 1971.
6. Woolrich, P. F., "Occurrence of Trace Metals in the Environment—An Overview," *J. Am. Ind. Hyg. Assoc.* (1973) 34, 217.
7. "Criteria for a Recommended Standard . . . Occupational Exposure to Inorganic Arsenic New Criteria—1975," HEW Publication No. (NIOSH) 75-149, U.S. Department of HEW, Public Health Service, Center for Disease Control, NIOSH, Ohio, 1975.
8. "Criteria for a Recommended Standard . . . Occupational Exposure to Beryllium," Publication Number HSM 72-10268, U.S. Department of HEW, Public Health Service, Health Services and Mental Health Administration, NIOSH, Ohio, 1972.
9. "Criteria for a Recommended Standard . . . Occupational Exposure to Cadmium," HEW Publication Number (NIOSH) 76-192, U.S. Department of HEW, Public Health Service, Center for Disease Control, NIOSH, Ohio, 1976.
10. "Criteria for a Recommended Standard . . . Occupational Exposure to Chromium (VI)," HEW Publication Number (NIOSH) 76-129, U.S. Department of HEW, Public Health Service, Center for Disease Control, NIOSH, Ohio, 1975.
11. "Criteria for a Recommended Standard . . . Occupational Exposure to Inorganic Lead," Publication Number HSM 73-11010, U.S. Department of HEW, Public Health Service, Health Services and Mental Health Administration, NIOSH, Ohio, 1972.
12. "Criteria for a Recommended Standard . . . Occupational Exposure to Inorganic Mercury," Publication Number HSM 73-11024, U.S. Department of HEW, Public Health Service, Health Services and Mental Health Administration, NIOSH, Ohio, 1973.
13. "Criteria for a Recommended Standard . . . Occupational Exposure to Inorganic Nickel," DHEW (NIOSH) Publication Number 77-164, U.S. Department of HEW, Public Health Service, Center for Disease Control, NIOSH, Ohio, 1977.
14. Baier, E. J., Deputy Director, NIOSH, statement at public hearings on occupational standard for beryllium, Occupational Safety and Health Administration, Department of Labor, Washington, D.C., 1977.
15. Baier, E. J., Deputy Director, NIOSH, statement at public hearings on occupational standard for lead, Occupational Safety and Health Administration, Department of Labor, Washington, D.C., 1977.
16. "Recommended Standard for Occupational Exposure to Vinyl Chloride," U.S. Department of HEW, Public Health Service, Center for Disease Control, NIOSH, Ohio, 1974.

RECEIVED December 12, 1977.

Analysis of Molybdenum in Biological Materials

GLENN E. BENTLEY, LAURI MARKOWITZ,
and ROBERT R. MEGLEN[1]

Environmental Trace Substances Research Program, University of Colorado,
Boulder, CO 80309

A procedure for the determination of molybdenum in serum, red blood cells, and urine is described. The low concentration of molybdenum in most unexposed individuals requires the sensitivity obtained using atomic absorption spectrophotometry and electrothermal atomization. Spike recovery tests indicate that low temperature ashing is required for accuracy. Severe matrix interferences preclude wet ashing or high-temperature ashing as sample pretreatments. Using the method described, it is possible to distinguish between industrially exposed and unexposed individuals.

In recent years there has been an increased interest in assessing the human health effects from environmental exposure to trace metals. Studies of occupational exposures and dietary intakes of trace metals have required the refinement and development of analytical techniques for the analyses of low elemental concentrations in complex matrices. Molybdenum is one of the trace metals that has been the subject of intensive study because it is an essential trace element in both plant and animal nutrition. It is an integral constituent of several metalloenzymes including xanthine oxidase, which is the last enzyme in the catabolic pathway of purines. Extensive ingestion of molybdenum has been shown to cause molybdenosis in cattle. Many of the features of this condition can be ascribed to induced copper deficiency. Whether biochemical changes or adverse health effects in humans can be attributed to excessive exposure to molybdenum is not known.

[1] Author to whom correspondence should be addressed.

0-8412-0416-0/79/33-172-033$05.00/0

More than 15,000 analyses of waters, foods, plants, soils, and biological tissues have been analyzed as part of a research project on health effects of molybdenum. Results of this work will be used to prepare a drinking water criteria document for the U.S. Environmental Protection Agency.

Biological fluids such as serum or plasma, red blood cells, and urine are particularly difficult to analyze. The low molybdenum concentrations found in normal human samples are below the detection limit of the thiocyanate colorimetric method (100 ng) and much below conventional flame absorption spectroscopy (1 μg). Normal blood levels of molybdenum are about 10 μg/L and sample volume is usually $\leq$ 1 mL. The low concentration and limited sample size preclude direct analysis or sample preconcentration for analysis by the conventional analytical methods.

The graphite furnace technique for atomic absorption spectrophotometry permits molybdenum determination at the ng level with small sample volumes. This technique is, however, prone to matrix interferences when used for molybdenum analysis. The boiling point of molybdenum is about 4600°C, about 1800°C above the maximum temperature obtained in the graphite furnace. Therefore, molybdenum atomization can only take place by a mechanism which includes the formation of a compound that is volatile at 2800°C. Molybdenum atoms will only be produced if the volatile molybdenum compound dissociates at this temperature. Any ions or compounds which affect the complicated atomization mechanism will alter the sensitivity of the method. This crucial fact is the primary reason for the extreme matrix sensivity of the method.

Because of the matrix sensitivity, blood, serum, and urine samples must be treated to destroy organic constituents prior to analysis. Inorganic residues from these samples also have high concentrations of several potential interferents. Dry ashing techniques were investigated because wet ashing methods requiring sulfuric or perchloric acids further add to the potential interferences by increasing the ionic content of the samples.

The signal suppressing effects of biological matrices preclude the analysis of the samples by using aqueous calibration standards. Matrix duplication for the standards proved unreliable. Removal of the interfering sample matrix by chelating ion exchange resins was not reproducible at molybdenum levels below 0.1 μg. Therefore, the method of standard additions has been used for the analysis of these complex materials. This method of calibration is time consuming but has been used successfully with these samples.

Experimental

Apparatus. All atomic absorption measurements were made with a Perkin–Elmer Model 360 Atomic Absorption Spectrophotometer equipped with a Perkin–Elmer HGA 2100 Graphite Furnace. The spectrophotometer conditions were: slit: alternate, 0.7 nm; mode: time constant 1, expansion: 15×. A Perkin–Elmer molybdenum hollow cathode lamp was operated at 30 mA. The monochromator was set to pass the 313.3 nm resonance line of molybdenum. The graphite furnace conditions were; dry: 50 sec at 120°C; char: 40 sec at 1800°C; atomize: 8 sec at 2800°C. The argon internal purge gas was set at continuous flow (normal) at 30 mL/min.

Sample Preparation. SERUM: A 0.3–1.0 mL aliquot of each serum sample was transferred to a 2 mL borosilicate glass vial with a micro pipet. Three drops of concentrated nitric acid were added and the samples were dried overnight at 70°C. The dried samples were then placed in a Tracerlab Model 600 Low-Temperature Asher (LTA) for 8–24 hr. When the ash was almost white, one drop of concentrated nitric acid was added to each sample. The samples were again dried at 70°C and returned to the LTA for an additional 8–24 hr. The ash residues were taken up to their original volume with 1:1 aqueous hydrochloric acid. The dilute acid was added in such a way to ensure that any residue adhering to the sides of the vial was in contact with the diluent. At least 0.5 hr was necessary for complete dissolution of the sample. RED BLOOD CELLS (RBC): For clinical procedures, the results of the RBC analyses were reported on a volume basis. For these samples, disposable pipet tips were modified by cutting about 2 mm off of the tip to widen the opening. Each aliquot of RBCs was then transferred to a vial. The pipet tip was rinsed with three successive volumes of deionized water. These washes were added to the sample vial together with three drops of concentrated nitric acid. The samples were predigested at room temperature for 12–16 hr. When the samples appeared gray-brown in color the vials were placed in a drying oven at 70°C for 12 hr or until dry. The dry samples were transferred to the LTA and ashed for 16 hr. Additional treatments with one drop of concentrated nitric acid, drying, and ashing were repeated until the ashes appeared white. The sample was dissolved in an equal volume of 1:1 aqueous nitric acid. URINE: Urine sample preparation and dissolution was the same as for the serum.

Procedure. Analyses were performed using the graphite furnace conditions described in the apparatus sections. Because there is a severe suppression of the molybdenum absorption signal, it was necessary to use the method of standard additions. Standards containing 0, 10, 20, 30, and 50 μg molybdenum per liter were prepared from a stock standard solution of molybdate (prepared from MoO_3). Background correction is not used in this procedure (*see* Discussion section); however, the furnace blank is sufficiently high to require that a blank correction be made. Several shots with no sample or standard were made to determine the correction. Twenty-five μL of the sample were injected into the furnace and the molybdenum absorbance was recorded with a strip chart recorder. Standard additions were made to the furnace using the standards described above. In each case, a 25 μL aliquot of the standard preceeded the injection of the 25 μL of sample. The furnace program was then initiated and the absorbance was recorded. A high temperature burn, at furnace maximum, was performed after each signal to reduce memory effects. An additional furnace blank determination followed each set of additions. Pyrolytically coated graphite tubes obtained from the furnace manufacturer were used for all determinations. Linear regressions were performed to determine the concentration of molybdenum in the sample.

Results and Discussion

A series of spike recovery tests indicated that poor recovery was obtained on samples that were ashed in a muffle furnace at 40°C. Therefore, low-temperature ashing (LTA) was used on all samples. Table I shows the results of these tests. The average recovery of six molybdenum-spiked samples showed only 70% recovery after muffle-furnace ashing at 450°C. The low recovery is not completely understood. Similar spike recovery tests on plant material and animal tissues do not show this apparent "loss." Our data suggest that the blood and serum matrix is different from other biological materials. The molybdenum in the spikes may not be lost through volatilization but instead may be chemically converted to a form that is not easily dissolved. Another possible explanation is that the molybdenum is converted to a species that is not readily atomized in the graphite furnace.

Background correction for molecular absorption and light scattering at the 313 nm molybdenum resonance line leads to unacceptable noise levels at the large scale expansions used. Spike tests at higher than normal molybdenum concentrations indicate that background correction is not necessary when a high char temperature is used. Nonspecific absorption was determined to be less than the noise level. However, it is necessary to make a furnace blank correction. The furnace blank is determined by firing the furnace with no sample injection before and after each set of standard additions. The furnace blanks determined in this way are very reproducible.

Figure 1 shows a standard curve for aqueous molybdate standards (circles). The line through the points that are indicated by triangles is a standard additions experiment on a serum sample containing about 5 µg/L Mo. The serum matrix interference reduces the sensitivity markedly. The difference in slopes between the aqueous standards and the sample illustrates the matrix suppression effect. Figures 2 and 3 show

Table I.   Spike Recovery Tests

	Mo Added µg/L	Mo Recovered µg/L	%
LTA serum	10	11.2	112
	20	18.6	93
	30	29.1	97
	40	40.4	101
Muffle-ashed serum (average of six samples)			70

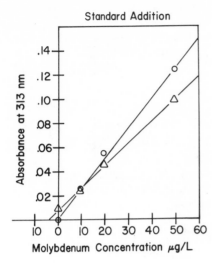

Figure 1. *Standard curve showing the absorption signal for aqueous standards (circles) and signals obtained for standard additions to ashed serum (triangles)*

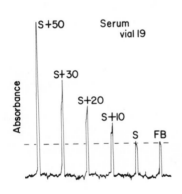

Figure 2. *Absorption signals obtained for standard additions to serum. Twenty-five μL of aqueous standards (concentration units are μg/L). Furnace blank is labeled FB.*

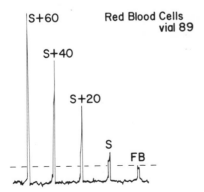

Figure 3. *Absorption signals obtained for standard additions to red blood cells. Twenty-five μL sample and 25 μL of aqueous standards (concentration units are μg/L). Furnace blank is labeled FB.*

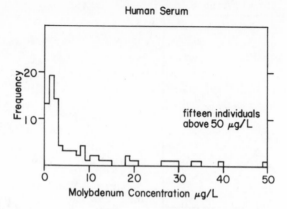

*Figure 4. Frequency vs. concentration of molybde-
num in serum for 80 individuals. Fifteen individuals
above 50 μg/L are not shown.*

the signals obtained from standard additions on serum and red blood
cells respectively. Figure 4 shows the concentration of molybdenum in
serum for eighty "normal" and industrially exposed individuals. All of
the normal levels lie between 0–30 μg/L. This figure is provided only to
indicate that it is possible to distinguish between exposed and unexposed
individuals. More complete data on health effects and exposure levels
will be discussed in other publications.

There is currently no standard reference blood or serum. It is therefore difficult to assess the accuracy of this method. However, analyses
on NBS Orchard Leaves and Bovine Liver are shown in Table II. For
these standard reference materials, both low-temperature ashing and
muffle-furnace ashing give results that agree within experimental error.
However, muffle-furnace ashing is not acceptable for serum samples.
There is a significant difference between the serum matrix and the standard reference materials. Therefore, care should be exercised in assessing
the accuracy of a method on the basis of standard reference materials
that differ from the sample matrix under investigation.

Table II.   Accuracy Checks

NBS Standard Reference Materials	*NBS* μg/g	*This Work* μg/g
1577 Bovine Liver	(3.2)[1]	3.1
1571 Orchard Leaves	0.3 ± .1	0.2
Both LTA and muffle ashing acceptable		

[1] Not yet certified.

In summary, a method for the analysis of molybdenum in biological fluids has been presented. The method requires the destruction of the organic materials in the sample by low-temperature ashing. Detection was accomplished by using a graphite furnace—atomic absorption technique and the standard additions method. The method is sufficiently sensitive to distinguish between molybdenum levels in the blood, serum, and urine from exposed and unexposed individuals.

Acknowledgment

The authors gratefuly acknowledge the financial support of the U.S. Environmental Protection Agency under Grant Number R-803645-02. We also wish to thank Glenda Swanson for technical assistance in developing the methods described and for helpful suggestions regarding the manuscript.

5

Analysis of Size-Fractionated and Time-Dependent Particulate Matter for Metals by Atomic Absorption Spectrometry or by Neutron Activation Analysis

G. J. ROSENBERGER, W. D. SMITH, W. W. MILLER, and T. H. RISBY [1]

Department of Chemistry, The Pennsylvania State University, University Park, PA 16802

This study confirms previous studies by other researchers that metals with anthropogenic origins are found in sub-micrometer-sized particles and that their emissions are periodic during a 24-hr period. These data were collected during 18 months in rural central Pennsylvania and representative analyses are presented. It is expected that these results are typical background levels present in similar geographical locations. Two analytical chemistry techniques were selected as representative of nondestructive and destructive methodologies. No attempts were made to collect and analyze duplicate samples by both methodologies although the results obtained on different days were comparable. Twelve elements were detected and analyzed by neutron activation analysis and eight elements were analyzed and four were detected by atomic absorption spectrometry.

Airborne particulate matter, which is formed by the condensation of gases or vapors or by mechanical or communitive processes, has been shown by many researchers to affect both flora and fauna. The two extremes of these affects are the accumulation of inert materials and of toxic or carcinogenic materials which lead to the eventual death of the biological specimen. As a result of such studies, society has attempted to reduce the total atmospheric burden by the enaction of emission control

[1] Current address: Department of Environmental Health Services, Johns Hopkins School of Hygiene and Public Health, 615 N. Wolf St., Baltimore, MD 21205.

legislation. These laws, by definition, require that the air is monitored, and in the United States of America there exist both state and federal sampling networks. Most of these data have been collected with high-volume filter samplers which give information on the gross loading of the ambient air (TSP), and for most purposes this information is sufficient. Although the majority of the mass of airborne particulate matter is nonmetallic (*1*), this study is concerned with the measurement of metals in particulate matter since they, by definition, are not biodegradable and therefore persist in the environment (*2*).

In addition to determining the gross concentration of airborne particles to establish the influence of anthrogenic sources, studies have been initiated to determine the relationship between particle size and chemical composition. These studies have shown, as a general rule, that larger particles (> 2.0 μm) often result from mechanical or communitive processes while smaller particles (< 2.0 μm) are formed by the condensation of gases or vapors. This general statement is substantiated by the data which have shown that the chemical composition of the gross and fine particles differ considerably, which implies different origins. Also the appearance of fine particles is generally an indication of human activity (4).

During the past 10 years many studies have been made concerning the influence that particle size can have on the health effects of airborne particulate matter. The size of the particle has a direct relationship on where it impacts on the respiratory tract of humans and animals. However, previous studies (*3*) on particle deposition in the respiratory system are not definitive although, as a general rule, large particles (> 1 μm) and fine particles (< 0.01 μm) are deposited in the nasopharyngal system, and particles between these extremes are deposited in the trachiobronchial and pulmonary systems. Particles that are deposited in the nasopharyngal and trachiobronchial systems are removed by mucus and enter the gastrointestinal tract within a period of about one day. Once in the gastrointestinal tract, particles can have a primary toxic effect. Particles that enter the pulmonary system are cleared more slowly (residence times up to one year) and therefore are more likely to be solubilized and to enter the blood stream. Particles deposited in the trachiobronchial system also can enter the blood stream directly but, for such absorption, the particles must be very soluble. This requirement probably means that metal ions are not absorbed in this region since it is unlikely that the airborne metals will have sufficient solubility. The absorption of metals via the pulmonary system is probably more efficient than absorption via the gastrointestinal tract, which means that particles containing toxic metals in size ranges that could be deposited in the pulmonary system are more environmentally significant. These health effects studies, coupled with the information published by Brosett et al. (5) on the comparative life times of gross vs.

fine particles in the atmosphere, have spurred research on the relationship between chemical composition and particle size. Major efforts in this area have been initiated by the following groups: California Tri-City Project (6), U.S. Environmental Protection Agency Cascade Impactor Network (7), and ACHEX Study (8, 9), and while the results are still preliminary, these studies are steps in the right direction. There have been also a number of studies by individual researchers. Wesolowski et al. (10) have reported the use of a Lundgren impactor to size-fractionate particles in San Jose and have measured the variations of vanadium, bromine, sodium, and aluminum concentrations with size. This study showed that the sources of vanadium and bromine were anthropogenic whereas the origins of sodium and aluminum were natural. Other studies also have shown this general trend: high concentration of lead in fines (11); high concentrations of metals from combustions sources in fines and high concentrations of metals from soil dusts, etc. in coarse particles (12, 13); in the ambient air of rural areas there are low concentrations of anthropogenic metals, which is exactly the reverse of what is found in urban areas that contain high concentrations of lead, iron, copper, manganese, nickel, vanadium, and zinc (14); in St. Louis and surrounding areas, high concentrations of lead and sulfur have been found in fines (15, 16) as opposed to those found in coastal areas where the metals are found in coarse particles (18). In another study, Schuetzle et al. (20) have used high resolution mass spectrometry to show how both anthropogenic sources of metals and organics were preferentially concentrated in fines. Other studies have been made in this area but these examples are illustrative of the published work. This idea that the sources of particles designates size has been well established by previous studies and also that the diameters of airborne particles follow a bimodal distribution, although a recent report on emissions from catalyst-equipped cars has shown that these emissions have a trimodal distribution (21). Therefore, Dzubay and Stevens (22) designed a two-stage dichotomous filter that traps the anthropogenic and natural particles separately. This study showed that zinc, sulfur, bromine, arsenic, selenium, and lead were contained in fines and silicon, calcium, titanium, and iron in coarse particles. While this study provides a method of separating anthropogenic and natural airborne particles and also provides useful information on the gross sizes of airborne particles, further studies which measure chemical composition as a function of particle size are required.

Another area which also warrants interest is the measurement of the time dependence of particulate emissions. At this time data are available on the daily or monthly variations of particulate emissions, which is the first step in this direction. However, there have been no studies which show any temporal relationship of particulate emissions. Many publications and observations have suggested that emissions vary during the day and often increase significantly at night. The verification of these trends will be important since it will then be possible to identify which segment of the population is exposed during peak emissions. Exposure is a function of dosage and time, and it is possible that exposure to high concentrations over short periods of time can present a serious health hazard.

In addition to the obvious environmental effect of airborne particulate matter, there have been a number of studies which have shown that particles have direct effects on the climate. Particles can scatter and absorb solar radiation, change levels of precipitation, cause the formation of warm or cold clouds, and cause the formation of fogs or smogs.

This chapter presents data on the chemical composition of size-fractionated particles and the temporal distribution of metallic emissions. These data were collected at The Pennsylvania State University which is situated at University Park in central Pennsylvania. The location is rural and is at least 100 miles from any major industrial source or any major urban area. It was expected that the major concentration of particles would result from natural sources such as entrained soil although some contributions from anthropogenic sources, such as the combustion of fuels for heating and transportation, were expected.

The sampling and analysis of airborne particles presents major problems to analytical chemists since, by definition, the ambient air is heterogeneous and it is seldom possible to obtain a representative sample. Also, as the volume of air sampled is decreased the problem of representation is enhanced. This problem is magnified by the sampling site since the total atmospheric loading in central Pennsylvania is low and therefore the sample can have metallic concentrations that are not significantly higher than the blank. Therefore it was essential that trace analytical chemistry methodologies and sampling procedures with low metallic impurities be adopted. Many analytical chemistry techniques are suitable for these analyses (23), and neutron activation analysis and flame atomic absorption spectrometry were chosen for this study. The former methodology was chosen since it is a multi-element nondestructive technique and it requires a minimum amount of sample preparation. The latter advantage is extremely important since many of the particles consist of metallic compounds contained in silica or silicate matrices which are only solubilized with great difficulty. Atomic absorption spectrometry was chosen since it is a popular selective trace method of analysis. A flame was used as the atom reservoir instead of a nonflame reservoir as a result of difficulties experienced in prior studies in this laboratory (24) in which irreproducible atomization was observed for metals contained in matrices with low atomization efficiencies. The disadvantages of using the flame are that the limit of detection is poorer and that larger quantities of solution are required.

Sampling procedures for the size fractionation of airborne particles use devices that aerodynamically size particles according to their inertial mass. These devices typically consist of fixed or rotating plates that are designed to change the course of the air stream. This change of direction causes those particles with too large an inertial mass to leave the airstream and impact on the collection device. This process is repeated and a series of collected size ranges is obtained. The size distribution in each range depends upon the design of the collection device and the distribution of the densities of the various particles. The major problems experienced with such devices are that the particles can bounce off the impaction surfaces and impact at another surface or can adhere to the walls of the collection device. These problems are difficult to evaluate and as a result, although theoretically the size distributions can be calculated for each impaction surface from aerodynamic considerations, the experimental distribution is much broader. There are a number of commercial devices available that can size fractionate particles and collect them in a form suitable for subsequent chemical analysis. These devices include the Lundgren impactor, the Anderson impactor, the Casella impactor, and the Unico impactor. The Lundgren impactor uses a high linear gas velocity whereas the other three use medium-to-low linear gas velocities. The Anderson impactor has the advantage that it has 6–8 stages, which means that a much greater size differentiation is obtained. All of these impactors can be modified so that membrane filters can be used as the impaction surfaces, which makes subsequent chemical analysis easier since the background contamination is low. The temporal collection of airborne particles can be performed with membrane filters, as size fraction is not possible since the atmospheric loading is low.

Experimental

Apparatus. 1. NEUTRON ACTIVATION ANALYSIS. The neutron activation analysis was performed at The Pennsylvania State University TRIGA reactor with a thermal neutron flux of about 1×10^{13} neutrons/cm^2 sec. A 36-cm^3 co-axially drifted germanium (lithium) semiconductor detector was used to monitor the gamma rays. This detector is hermetically sealed and maintained at liquid nitrogen temperatures to minimize noise and ion drift. The detector has a bandwidth at half-height of 2.0 keV at the 1332 keV gamma peak of ^{60}Co. The signal is fed into a 1024-channel analyzer (ND-2200, Nuclear Data Corp.). The amplifier was calibrated over a working range of 0–2 MeV, using as references the ^{241}Am, ^{60}Co, and ^{22}Na nuclides. The digital output from the multichannel analyzer is manipulated using the master program ENERGY (25) and additional subroutine algorithms.

2. ATOMIC ABSORPTION SPECTROMETRY. The atomic absorption spec-

trometric analyses were accomplished using either an air–acetylene or a nitrous oxide–acetylene flame (Varian-Techtron Model 1000).

Procedure. 1. AIR SAMPLING. The size-fractionated particles were collected with a four-stage cascade impactor (Unico Corp.) and a 25-mm 0.2 μm back-up filter (Nucleopore Corp.). The selection of this impactor was based solely on availability. Ambient air was drawn through the impactor with a low-volume pump at a flow rate ca. 0.018 m³/min. The system also included a calibrated flow meter and manometer. The impactor is designed to use microscope slides as the impaction surface. These present difficulties in subsequent analysis of the collected particulate matter in that the slides contain impurities, particularly sodium, which will interfere with the neutron activation analysis. Therefore membrane discs (36-mm 0.1 μm Nucleopore Corp.) were placed at the entrance to each jet and these discs were backed by the glass slides. Prior to sampling, the glass slides were cleaned by immersion in the following solvents: alcoholic potassium hydroxide ($1M$), de-ionized water, nitric acid ($16M$ Ultrex, J. T. Baker Co.), and de-ionized water. After cleaning, the slides and filters were handled with Teflon forceps.

The time-dependent particles were collected on membrane filters (47 mm, 0.1 and 1.0 μm, Nucleopore Corp.) contained in polycarbonate filter holders (Nucleopore Corp.). The filter holders were also cleaned in a similar manner as the glass slides to prevent cross contamination from previous sampling. As with the impactor studies, the sampling train included a low-volume pump, flow meter, and manometer. This enabled accurate measurements of the flow-rate of ambient air to be made ($\simeq 0.018$ m³/min).

2. IMPACTOR CHARACTERISTICS. The impactor characteristics can be calculated if the densities of the particles are known. However, since this study was only for the size-fractionation of particles, the distribution was measured by optical microscopy. Ambient air was sampled for 1 hr and the particles were collected on glass slides. The particles were matched to the size circle on the calibrated Porton graticule. Approximately 100 particles were counted on each stage and the size distributions were generated.

The aerodynamic size distributions of particles contained on each stage of the impactor can be calculated providing that the densities of the particles are known. However, it is reasonable to expect that there will be a distribution of particle densities as a result of sample origin, which tends to broaden the particle distribution. This empirical calculation was checked experimentally.

3. ANALYSIS OF PARTICLES BY NEUTRON ACTIVATION ANALYSIS. The analytical neutron activation procedure used for the collected particles involved two irradiations of each filter (short-and long-term) using the standard procedures. The standards for neutron activation analysis were made with aliquots of known standards (1.000 ppt Fisher Scientific Standards) in sealed tubes. These standards were always counted with the samples. The following were the standards used in this study: Na and Cl; Cu and Mn; Al, V, and Ti; K and Br; Zn and Hg; Co and Cr; As and Ag; Se, Sc, Sb, and La; and Fe, Mg, and I.

4. ANALYSIS OF PARTICLES BY ATOMIC ABSORPTION SPECTROMETRY. Various methods of sample preparation were attempted to solubilize the

collected particles. The first approach was to ash the filter and particulate material in a reduced pressure (0.05 Torr) rf electrical discharge in oxygen (low-temperature asher, Tracor LTA 600). After the filter was completely ashed (24 hr), the ash was dissolved and diluted to a known volume (50.0 mL) with nitric acid (1.6M, Ultrex). Another procedure was to place the membrane filter in a previously cleaned Teflon beaker with a known volume (5.00 mL) of nitric acid .(16M, Ultrex). This beaker was then placed in an ultrasonic bath for 30 min. The temperature of the bath was maintained at 25°C to minimize any sample losses. After this time, the filter was removed with Teflon forceps washed with distilled water to a known volume (50.0 mL). The final method of preparation used a mixed solvent system consisting of chloroform (1.00 mL) and nitric acid (3.0M, 1.00 mL). The rolled membrane filter was placed in a previously cleaned glass ampule, the solvent system was added, and the resulting solution was refluxed. Refluxing was maintained by wrapping the top of the ampule with copper tubing (3 mm o.d.) through which cooling water was passed. After 1 hr of reflux the aqueous layer was separated and diluted to volume (10.0 mL). In another procedure the filter was placed in the reflux ampule with an aliquot (1.00 mL) of nitric acid (1.6M). The resulting solution was heated in an ultrasonic bath for 30 min. After this time the solution was transferred to a volumetric flask (10.0 mL) and diluted to volume with nitric acid (1.6M) that had been used to rinse the reflux ampule and filter. All of the standards (Fisher Scientific Standards) were made in nitric acid (1.6M), and all of the glassware was cleaned with alcoholic potassium hydroxide (1M), followed by nitric acid (16M). The most suitable flame for the production of ground-state metal atoms for each element was used, and the concentration was determined using a standard calibration curve. The method of standard additions was not used since no interferences from the matrix were observed.

Results and Discussion

(1) **Cascade Impactor Characteristics.** If the cascade impactor is to be used to obtain information on the diameters of particles in the atmosphere, then the distribution of the particle sizes on each stage must be known. Two approaches were used to obtain these data. The first used an empirical equation derived for this impactor (26, 27, 28):

$$D_{50} = 2.5 \left(\frac{\eta w}{C \rho v_j} \right)^{1/2}$$

where D_{50} is the diameter (cm) of the particles that are impacted with 50% efficiency, η is the viscosity of ambient air ($\sim 1.84 \times 10^{-4}$ poise), w is the width of the impactor jet, ρ is the density of the particle (~ 1.5 g/cm^3), v_j is the jet velocity in cm/sec, and C is the Cunningham correction factor which, for particles greater than 1 μm can be shown to be 1 (stages 1–3) and for particles less than 1 μm, $C = 1 + 2.5 \times 10^{-6}/D_p$ (stage 4) (this factor is dimensionless and D_p is the particle diameter).

Under typical sampling conditions, the following aerodynamic diameters which are collected with 50% efficiency were calculated for each stage: Stage 1, 9.8 μm; Stage 2, 4.2 μm; Stage 3, 1.8 μm; Stage 4, 0.7 μm; and Back-up filter, < 0.7 μm. In the second approach the actual particle diameters were measured by optical microscopy. From this study the maximum number densities for each stage were found to be in the following ranges: Stage 1, 7.69–10.88 μm; Stage 2, 2.73–5.33 μm; Stage 3, 1.07–2.15 μm; and Stage 4, 0.65–1.85 μm.

(2) **Background Caused by Filters.** Since all of the particles were collected on membrane filters it was necessary to determine the blank metal concentrations in the filter. This enabled an estimation of how many particles must be collected in order that the levels of the metals were significantly greater than the blank filter. For this study, both neutron and flame atomic absorption spectrometric analyses were used and the results are shown in Table I. The analyses by neutron activation were made on the filter directly whereas those by atomic absorption spectrometry were obtained by extracting the filter with nitric acid (16M Ultrex). There are apparent differences between the two sizes of membrane filters which are probably related to the fact that these filter sets were obtained at different times. Also, while the metal blanks within a particular batch of filters vary by negligible amounts, the variations between batches are considerable. These determinations are near the detection limits for both techniques, and therefore there are considerable uncertainties associated with the results. However, these blanks did indicate the minimum level of metals which must be collected if the analyses are to be significant.

(3) **General Results from the Analytical Methodologies.** NEUTRON ACTIVATION ANALYSIS. The only problems experienced with this technique resulted from peak overlap and peaks which occurred on the continuum background. The peak overlaps that were observed are as follows: 617 keV ^{80}Br and 620 keV ^{38}Cl double escape peak; 844 keV ^{27}Mg and 847 keV ^{56}Mn; 554 keV ^{82}Br, 559 keV ^{76}As, and 564 keV ^{122}Sb; and 1116 keV ^{65}Zn and 1120 keV ^{46}Sc. These peaks were recalculated since the program will often only detect one large peak. The peaks that occurred on the continuum were also recalculated to subtract the background contribution.

ATOMIC ABSORPTION SPECTROMETRY. Flame atomic absorption spectrometry was adopted as the second method of analysis and since low volumes of air were sampled, only a limited number of elements were detected in the collected particles (calcium, copper, iron, magnesium, and zinc). Aluminum, cadmium, chromium, cobalt, lead, manganese, and nickel were not detected in any of the samples. This resulted in a limited

Table I.   Elemental Levels (ng) in
Neutron Activation Analysis

Element	36-mm Filter	Number Detected /15	47-mm Filter	Number Detected /15
Sodium	3890 ± 55	15	264 ± 58	15
Magnesium	< 100	0	< 100	0
Manganese	2.85 ± 0.57	15	4.24 ± 3.0	13
Bromine	13.2 ± 2.17	13	11.7 ± 5.5	12
Aluminum	7.36 ± 1.52	13	107 ± 44.3	15
Chromium	279 ± 62	15	32.5 ± 6.0	15
Zinc	150 ± 70	12	94 ± 32	15
Cobalt	4.77 ± 2.50	10	5.0	5
Iron	60 ± 150	10	510	5
Mercury	31.0 ± 18.0	15	21.3 ± 12.0	15
Vanadium	< 2	0	< 2	0
Titanium	< 100	0	150	7
Copper	52	5	< 100	0
Scandium	0.20 ± 0.12	12	0.30 ± 0.10	13
Tin	< 100	10	100	0
Antimony	< 10	0	1.4	5

[a] These results were obtained with chemically cleaned filters in a previous study in our laboratory (26).

data set; however, if the location of the sampling site was changed from a rural to an urban center, then more elements would have been detected.

The results of the various methods used for the dissolution of the collected particulates are shown in Table II. The final method was the easiest way of producing a homogeneous solution of the particulate matter.

COLLECTION SITE. All of the samples were collected either on the University Park campus of The Pennsylvania State University or in the surrounding community. Air was sampled only on days that were preceeded by at least 48 hr without precipitation to ensure that the particle distribution was typical.

(4) **Analysis of Size-Fractionated Particles.** A series of samples was analyzed by both methods of analysis, and the results of two samples that were collected on different days are shown in Tables III and IV. The data shown in Table III illustrate that there are probably two origins of the elements. Vanadium, bromine, manganese, copper, mercury, chromium, and zinc increase in concentration as the particle size decreases. This inverse relationship is expected if these particles are emitted by high-temperature combustion processes such as automobiles and power plants (which are the major sources in this area). Sodium, aluminum, iron, scandium, and cobalt were present in an approximately uniform distribution throughout the particle size range. This relationship results

Blank Filters (0.1 μm, Nucleopore)

Atomic Absorption Analysis

Flame 47-mm Filter	Number Detected /10	Non-Flame[a] 47-mm Filter	Number Detected /15
NA[c]		NA[c]	
108	10	94	15
ND[b]	0	11	15
NA[c]		NA[c]	
ND[b]	0	640	15
ND[b]	0	66	15
220	10	35	15
ND[b]	0	NA[c]	
860	10	350	15
NA[c]		NA[c]	
NA[c]		NA[c]	
NA[c]		NA[c]	
ND[b]	0	NA[c]	
NA[c]		NA[c]	
NA[c]		NA[c]	
NA[c]		NA[c]	

[b] ND—not detected.
[c] NA—not analyzed.

from the diminution of soils (*20, 21*) and can therefore be expected to be uniform in size distribution. These observations were checked to see if the metals originate from natural or anthropogenic sources by calculating the enrichment factor. Enrichment factors for these elements were approximately equal to the natural abundance of these elements in soils on days when the ambient temperature was warm, but on cold days there were contributions from other sources (probably combustion of fossil

Table II. Dissolution of Collected Particulate Matter

Method	*Observations*
Low temperature ashing and dissolution with nitric acid.	At least 24 hr were needed for complete ashing of the filter.
Ultrasonic bath at 25°C with nitric acid.	Incomplete dissolution of the particles.
Refluxing with chloroform and nitric acid in a reflux ampule.	The membrane filter once dissolved in chloroform partitioned with the aqueous phase and caused an increase in the absorbance measurements via molecular absorption.
Ultrasonic bath at 100°C with nitric acid in a reflux ampule.	Complete dissolution of the particles.

Table III. Analysis of Size-Fractionated

Total Metal Collected (%)

Stage	$D_{50}(\mu m)$	Na	Al	Br	Mn	Cu	V
1	9.8	22.6	14.8	7.0	3.5	12.5	4.7
2	4.2	28.3	23.3	6.8	11.3	16.0	0
3	1.8	1.3	25.5	18.3	25.4	49.9	30.6
4	0.7	27.8	36.4	67.8	59.9	21.6[a]	64.7

Total Atmospheric Concentration ($\mu g/m^3$)

		0.318	0.089	0.013	0.007	0.015	0.004

[a] Four-hour sample, 0.017 m³/min, 11/20/77.

Table IV. Analysis of Size-Fractionated Particles[a] by Flame Atomic Absorption Spectrometry

Stage	$D_{50}(\mu m)$	Total Metal Collected (%)			Total Atmospheric Concentration ($\mu g/m^3$)		
		Fe	Mg	Zn	Fe	Mg	Zn
1	9.8	20.0	27.1	11.4	.098	.080	0.28
2	4.2	14.5	25.2	14.2			
3	1.8	36.5	26.5	15.3			
4	0.7	21.0	19.3	19.9			
Back-up filter	< 0.7	8.0	1.9	39.2			

[a] 24-hr sample, 0.012 m³/min, 5/11/77.

Table V. Analysis of the Variations of Metal Emissions during

Element	Time Period		
	14-20 00	20-02 00	02-09 00
Na	0.43	0.34	0.29
Al	0.21	0.32	0.25
Mn	0.009	0.012	0.014
V	0.007	0.006	0.006
Br	0.028	0.022	0.031
Ti	0.056	0.027	0.054
Cu	0.029	0.049	0.020
Hg	0.005	0.004	0.002
Zn	0.111	0.052	0.062
Cr	0.009	0.008	0.006
Fe	1.5	0.150	0.17
Co	0.003	0.002	0.002

[a] 0.1 μm filter 2/19/77 to 2/21/77.

Particles[a] by Neutron Activation Analysis

Total Metal Collected (%)

Ti	Fe	Zn	Co	Hg	Cr	Sc
15.4	43.2	26.5	33.3	11.7	0	6.9
29.5	0	0	0	12.2	0	0
23.1	0	12.9	0	27.8	24.1	37.9
32.0	56.8	60.6	66.7	39.1	75.9	55.2

Total Atmospheric Concentration ($\mu g/m^3$)

0.039	0.108	0.037	0.0002	0.003	0.008	0.0007

fuels). The data shown in Table IV is less conclusive since a smaller number of metals were detected, although iron and magnesium were probably the result of natural sources and zinc from man-made sources.

However, even with this limited data set it was possible to state that emissions from high-temperature sources are contained in particles which are in the (respirable) size range that can be trapped in the respiratory region. If the data contained in Table III and Table IV are compared for the elements that both techniques could detect (Fe and Zn), then the total atmospheric loading for Fe is 0.108 g/m³ by NAA and 0.098 g/m³ by AAS and for Zn is 0.037 g/m³ by NAA and 0.028 g/m³ by AAS. These samples were obtained on different days and yet the total loadings were comparable. However, the distribution of metals in the various particle

a 48-hr Period[a] by Neutron Activation Analysis ($\mu g/m^3$)

Time Period

09-15 00	15-22 00	22-07 00	07-13 00
0.45	0.48	0.28	0.32
0.42	0.30	0.33	0.18
0.020	0.012	0.005	ʼ0.006
0.009	0.014	0.002	0.005
0.022	0.094	0.017	0.024
0.063	0.071	0.061	0.031
0.052	0.051	0.011	0.035
0.002	0.004	0.002	0.003
0.063	0.081	ND	0.053
0.008	0.011	0.008	0.008
0.89	0.66	0.48	0.38
0.002	0.003	0.002	0.003

Table VI. Analysis of the Variation of Metal Emissions during a 24-hr Period by Flame Atomic Absorption Analysis ($\mu g/m^3$)

Date	Time of Day	Membrane Filter (μm)	Element Fe	Zn	Cu	Mg
2/23/77	1130-1600	0.1	.074	.088	.003	.066
2/23/77	1600-2100	0.1	.054	.049	.003	.080
2/23/77	2100-0930	0.1	.051	.058	.009	.048
3/3/77	1130-1600	1.0	ND[a]	.156	ND[a]	.115
3/3/77	1600-2100	1.0	ND[a]	ND[a]	ND[a]	.025
3/3/77	2100-0930	1.0	.972	.097	.052	.091
3/3/77	2300-0900	1.0	1.77	.011	.005	.242
3/8/77	1130-1630	1.0	.744	.045	ND[a]	.152
	(Series)	0.1	ND[a]	.009	ND[a]	.041
3/8/77	1700-2200	1.0	1.07	ND[a]	.013	.152
	(Series)	0.1	ND[a]	ND[a]	ND[a]	ND[a]

[a] ND—not detected.

sizes was markedly different; this could be the result of the neutron activation analysis sampling period being 4 hr vs. a 24-hr sampling period for the atomic absorption analysis samples. The more reasonable explanation is that since these samples were taken on different days, the size distribution of particles had changed. All samples were collected using the same sampling train so this could not be the result of instrumental differences.

(5) **Analysis of the Variation of Metal Emissions during a 24-Hour Period.** Some of the results of total particulate samples that were collected during 48-hr periods are shown in Tables V and VI. These data were obtained by neutron activation analysis and flame atomic absorption spectrometry. The data show that, in general, most elements reached maximum concentration in the afternoon or early evening period and were at a minimum in the early morning hours. The largest fluctuations are in those elements that are present primarily from high temperature combustion processes, i.e., bromine, vanadium, titanium, and zinc. The three periods of maximum bromine emission correlate with periods of heaviest traffic flow at 1700, 0700, and 1800 hours. The metals that result from natural sources showed much less time dependence.

Conclusions

This study has confirmed that metals from high temperature combustion sources are emitted in the form of respirable-sized particles and the air burden shows a definite time dependence. These results presented in this study are representative of a series of samples taken during the period May 1976–November 1977. Because of limitations in sampling equipment, it was not possible to obtain samples at the same time and subject them to both atomic absorption spectrometry and neutron activation analysis. However, attempts were made to first analyze samples by neutron activation analysis and then subsequently to analyze them by atomic absorption spectrometric analysis, but the results were very poor, probably as the result of contamination from the more extensive sample handling.

Literature Cited

1. Junge, C., *Ann. Meteorol.* (1952) **5**, 1.
2. Annon, *Chem. and Eng. News* (1971) **49**(29), 29.
3. "Airborne Particles," National Research Council, National Academy of Sciences, Washington, D.C., 1977, and references cited therein.
4. *Ibid*, page 69.
5. Brosset, C., Andreasson, K., Ferm, M., *Atmos. Environ.* (1975) **9**, 631.
6. Samuels, H. J., Twiss, S., Wong, E. W., Report of the California Tri-City Aerosol Sampling Project, Sacramento, California (1973).
7. Lee, R. E., Jr., *Science* (1972) **178**, 567.
8. Hidy, G. M., et al., Characterization of Aerosols in California (ACHEX) Final Report, Vol. I–IV, Air Resources Board, CA, 1974.
9. Hidy, G. M., Appel, B. R., Charlson, R. J., Clark, W. E., Friedlander, S. K., Hutchinson, D. H., Smith, T. B., Suder, J., Wesolowski, J. J., Whitby, K. T., *J. Air Pollut. Control Assoc.* (1975) **25**, 1106.
10. Wesolowski, J. J., Alcocer, A., Appel, B. R., AIHL Report Nos. 138-B, Berkeley, California, 1975.
11. Eisenbud, A. A., Kneip, T. J., Research Project 117, Electric Power Research Institute, 1975.
12. Dzubay, T. G., Stevens, R. K., *Environ. Sci. Technol.* (1975) **9**, 663.
13. Whitby, K. T., Lin, B. Y. H., P. L. Publ. Nos. 216, Minneapolis, University of Minnesota, 1973.
14. Altshuller, A. P., Characteristics of the Chemical Composition of the Fine Particulate Fraction in the Atmosphere, unpublished paper, May, 1976.
15. Dzubay, T. G., Stevens, R. K., ACS Environmental Chemistry, Vol. 16, 1 (1976).
16. Dzubay, T. G., Stevens, R. K., Peterson, C. M., p. 95 in X-Ray Fluorescence Analysis of Environmental Samples, Ann Arbor Science Publishers, 1977.
17. Winchester, J. W., *Bull. Am. Meteorol. Soc.* (1973) **54**, 94.
18. Winchester, J. W., Johanson, T. B., Grieken, R. E., *J. Rech. Atmos.* (1976) **8**, 761.
19. Natusch, D. F. S., Wallaco, J. R., Evans, C. A., Jr., *Science* (1976) **183**, 202.
20. Schuetzle, D., Crittenden, A. L., Charlson, R. J., *J. Air Pollut. Control Assoc.* (1973) **23**, 706.
21. Whitby, K. T., P. L. Report Nos. 253, University of Minnesota, Minnesota, 1975.
22. Dzubay, T. G., Stevens, R. K., *Environ. Sci. Technol.* (1973) **9**, 663.
23. Dulka, J. J., Risby, T. H., *Anal. Chem.* (1976) **48**, 640A.
24. Dulka, J. J., Risby, T. H., unpublished results, 1976.
25. Gunniuk, R., Levy, H. B., Niday, J. B., UCID—15140, Lawrence Livermore Laboratory, California.
26. Begnoche, B. C., Risby, T. H., *Anal. Chem.* (1975) **47**, 1041.

RECEIVED April 19, 1978.

6

Collection and Analysis of Airborne Metallic Elements

RICHARD J. THOMPSON

Analytical Chemistry Branch, Environmental Monitoring and Support
Laboratory, U. S. Environmental Protection Agency,
Research Triangle Park, NC 27711

The collection and analysis of airborne metal components encountered in air is categorized by physical state. Mercury, lead, and manganese are considered as metallic elements which can be found as components of ambient air. No examples are given for the liquid state. Attention is focused on particulate matter and its trace metal constituents—methods of evaluation considered include AAS, NAA, SS/MS, OES, XRF, and XRD. Elemental compositional levels and ranges of metals are considered. Fluctuations in the composition of samples taken at a site at differing times are noted; concentrational differences between sites can vary by 10⁵.

The Analytical Chemistry Branch (ACB) of the Environmental Monitoring and Support Laboratory of the U.S. Environmental Protection Agency has a number of responsibilities for analytical support. Analyses of fuels, sources, and ambient samples are performed along with the analyses of divers other types of specimens including tissue, both plant and animal. One of the major areas of support rendered by the ACB is in support of the National Air Surveillance Networks (NASN); a portion of this support consists of the analysis of collected material for airborne metallic elemental content. This chapter will, in the main, be a summary of the work done by the ACB with respect to the collection and analysis of airborne metallic elements.

Intuitively, one expects that trace metal analysis will involve collection and analysis of particulate matter. It is convenient to consider the collection and analysis of airborne species for the elemental determination of metals as a function of physical state—gas, liquid, and solid. In this chapter, ambient air analyses will be considered. Much interesting work has been done in source sampling and analysis; most of the data that have been evaluated for health effects have been obtained from industrial health studies of concentrated work room atmospheres. In the main, the analytical techniques applicable to samples taken from ambient air also are applicable to the examination of source and industrial health samples, although the sampling techniques that are used in source and industrial atmospheres (which contain 10^3 to 10^5 higher concentrations of metallic elements than does ambient air) are not amenable to use in ambient air without extensive modification.

Gaseous

Only substances inert to oxygen and to other atmospheric components which have significant vapor pressure would be expected to persist in ambient air. Only those substances which are introduced into air in vast amounts by man or nature would be expected to be found in ambient air. The only metal expected to be found in ambient air in elemental form and in the gaseous state is mercury. Mercury is an element which has other unusual characteristics besides that of the one of physical state. It forms covalent compounds of surprising viability and non-ionogens with some inorganic radicals. For example, mecuric cyanide is very soluble in diethyl ether. Mercury also is found in the atmosphere in a combined state and is not expected to be found in the combined state to any appreciable extent (less than 10% of the total mercury) in ambient air in locales other than those where organo-mercurials are produced or consumed. In the vicinity of a mine site, mercury levels as high as 20 μg of Hg/m^3 of sampled air are said to exist (1). Background levels are in the vicinity of 1 ng of elemental Hg/m^3 of ambient air (2). Obviously if one is to sample mercury in a chloro-alkali cell room or at a mine site, the technique applicable here would not necessarily be applicable to background monitoring because of the tremendous (10^5) differences in concentration as well as the difference expected in chemical form. Most of the mercury is in the elemental state in a chloro-alkali plant or in stack gases (3, 4). In ambient air, the predominante end product which one would expect from release into air of mercury species is elemental mercury.

A method has been proposed which is amenable for use for sampling of all atmospheres for mercury in all forms (5). The method as it has been used does not allow for the differentiation between the various forms of airborne mercury possible. Ambient air is drawn over silica wool held at a temperature of 400°C and then passed through silver wool by a pump. The combined species of mercury, if present, are pyrolyzed on the silica wool to elemental mercury, and all elemental mercury in the gas stream is collected by the silver wool. Should high concentrations of halogen species be present, they must be removed by absorption on a suitable collector after a pyrolytic step (Ascarite was used successfully) to avoid intereference by reaction with the silver wool. The silver wool collector is then placed into an analytical train, purged with air scrubbed of mercury and possible interferants, and then heated. The silver wool releases the captured mercury quantitatively at 400°C. The mercury so released is then determined by flameless atomic absorption by monitoring the light from a hollow cathode or other mercury lamp passed through a sample cell at 253.7 nm. The absorption of light by known amounts of mercury is measured and a calibration curve constructed; mercury present can be quantitated by relating instrumental response to the calibration curve. Obviously, for ambient air this system is amenable for use without the pyrolysis unit and by use of the silver wool collector alone, elemental mercury can be monitored per se in ambient air. With suitable splitting and valving devices in a gas-flow stream and multiple absorption tubes in the analytical train, an analytical range of 1–10^5 ng of Hg/m^3 of sampled air is made possible. With the pyrolysis system removed, this system has proven to be a useful monitoring tool for elemental mercury (6).

Two things are of interest to note here: (1) mercury concentrations might range at different sites from 1 to 10^5 ng of Hg/m^3 of sampled air, and (2) this method as developed is not a method in which great reliance could be placed unless the analytical portion of the overall measurement system is confirmed by phenomenological checks with alternative methods or with standard reference materials, which in this case do not exist. In this particular case, since the method of collection is highly selective, i.e., amalgamation with silver, the need for confirmation of the analytical method by phenomenologically independent techniques is not as demanding, but there is always the need for confirmation of any analytical technique. With respect to the gaseous state, one can note that with mercury no more than 10% of the total mercury is expected in combined forms.

No more than 10% of the total lead escapes collection by filtration (and therefore no more than 10% of the total lead is expected to be in a gaseous form). This was shown to be the case at an underground garage where alkyl lead compounds might be expected to be at a maximum in ambient air (7).

Manganese will become an element of concern as methylcyclopentadienyl manganese tricarbonyl (MMT) usage goes up. MMT is predicted to be in 30% of automobile gasoline by the end of 1978. A definitive study of MMT in atmospheres at, near, and removed from use sites has yet to be performed.

Liquid

Although the existence of other metal species in the gaseous state is possible, it is unlikely (because of reactivity) at other than a production or use site. Metallic elements are assumed to be either in a gaseous form (as organo metallics except for mercury) or in the form of particulate matter; material in the gas phase in air can be collected with particulate matter.

Solid

Background. Although metallic elements in particulate form can be collected by absorption (with impingement) in liquid media and by impaction, the most common method of collecting particulate matter containing metallic species is by filtration, with the hi-volume sampler (HI-VOL). One can note, however, that particles up to 100 μm are detectable in particulate matter collected with a HI-VOL whereas the elimination of particles above 20 μm would be desirable. In practice, the biggest use of the hi-volume sampler is for the reporting of "TSP" (total suspended particulate matter); a method for TSP and standards based on data obtained by this method have been promulgated (8).

A tare weight is obtained on a glass-fiber filter (that has a collection efficiency of 0.3 μm particles of over 99.9%) which is then used in the collection of particulate matter from ambient air. The loaded filter is reweighed, and difference in weight is ascribed to particulate matter. This weight is divided by the volume of sampled air, and the TSP is reported in $\mu g/m^3$. TSP values are used for: assessing long-term trends, determining the impact on human health and welfare, developing control strategies for guidelines for regulations, and enforcing purposes. Ambient averages were sought initially, shorter term data being a secondary interest.

Particulate Monitoring in the National Air Surveillance Network (9)

In 1953 the Public Health Service in cooperation with state and local departments in air-pollution control agencies set up air sampling stations in 17 communities. The Federal Air Pollution Research and Technical Assistance Act (Public Law 159, 84th Congress) became effective in June, 1955. The network expanded to become national in 1957, at which time about 110 urban and 51 non-urban stations in the 50 states, the District of Columbia, and the Commonwealth of Puerto Rico were operated on a continuing basis. Currently, the National Air Surveillance Network (NASN) includes some 270 stations (10% of which are non-urban) where particulate matter is collected on glass-fiber filters and TSP is determined by state and local agencies. As shown in Table I, the precision under controlled conditions can be quite good. This does not imply quantitative collection, however. The HI-VOL has been shown to collect 5-μm particles essentially quantitatively in a wind tunnel where the air velocity was 15 ft/sec; the efficiency drops with particle size and air velocity (10). Under routine conditions of monitoring TSP, the 95% confidence interval for precision was found to be 3.7–4.6% (11). The sampling is conducted from midnight to midnight, and for purposes of satisfying State Implementation Plans, is carried out at least every sixth day. Some 4000 high-volume samplers are in use nationwide for monitoring by state and local agencies.

For elemental analyses for trace metals, quarterly composites are constructed from strips from individual filters for analysis by optical emission spectrometer (OES) for survey purposes or for specific (and usually retrospective) analyses for a given element, atomic absorption (AAS), neutron activation (NAA), x-ray fluorescence (XRF), anodic stripping voltammetry (ASV), absorption spectrophotometry (AS), or some other alternative method. Spark-source mass spectrometry (SS/MS) is used for qualitative and semi-quantitative multi-elemental analyses. The original purpose for the network was to determine TSP, and the determination of organics (such as benzo(a)pyrene), nonmetal inorganics, and trace metals was undertaken because the samples existed. The data so produced are published as EPA reports, e.g., EPA-600/4-76-041, August 1976, Air Quality Data for Metals 1970 through 1974 from the National Air Surveillance Networks, available from the National Technical Information Service, Springfield, Virginia 22161.

Analyses of Particulate Matter. For mercury, as an example, analysis of particulate matter is a fruitless task since one expects that less than 10% of the total mercury (which is itself not very high in ambient air) is in a combined form and therefore a component of particulate matter. That is not to say that only 10% of the mercury will be collected, however, since the particulate matter can serve as a holding agent for mer-

Table I.   TSP Reproducibility (with HI-VOL Samplers Using Glass-Fiber Filters at 100 m³/hr) (23)

		TSP in $\mu g/m^3$ of Sampled Air	
Run	No. of Samplers	Range	AV
1	6	88–93	91
2	6	49–54	52
3	6	71–76	73
4	6	81–88	86
5	5	61–63	62
6	6	81–91	87

cury. (It is noted, for example, that mercury in ignited fly ash will accumulate if the fly ash is allowed to cool in a laboratory atmosphere rich in mercury). It has been implied that glass-fiber filters contain trace elements at levels which preclude their use for atmospheric monitoring purposes (12). This is not strictly true (13, 14). The content of many of the trace metals obtainable by NASN extraction from glass-fiber filters is at or below the detection limit (DL) of methods commonly used for many of the elements; only Fe, Mn, and Zn were above the OES DL in 1970 (13). The blank values are determined and, where measurable, compensated for in reporting the trace element constituency of particulate matter (13, 15). In the filters procured for monitoring use nationally in calendar year 1978, the only trace metal that can be obtained by mixed acid extraction from unused filters at a level more than 50 μg per 8 by 10 in. filter by the optical emission spectrometer technique is iron (16). For the NASN samples, it is true that the amounts of alkali and alkaline earth elements and of aluminum and titanium have been sufficient to pose matrix problems and must be compensated for (15). It can be shown, however, that it is possible for data to be obtained by OES and by AAS on hundreds of real samples for selected elements that are effectively from the same data population; data have been presented for lead (17).

In EPA's monitoring efforts, most techniques used have been checked by at least three phenomenologically different techniques in the analytical sequence. In the case of lead, for example, AS (dithizone), AAS, XRF, and OES gave data from the same population (15, 16). The choice of methodology to be used is thus managerial, not technical. In this particular example, one of the techniques (XRF) did not involve extraction of the sample. On a routine basis, trace element definitions are made with the OES for reasons of economy both in terms of sample used and laboratory costs. Obviously since several different kinds of analyses are carried out and there is a need for residual samples to be retained in the sample bank for retrospective purposes, sample economy is of vital importance. (It is of note that some 180,000 samples dating back to 1957, when metals analysis began, are currently stored in the sample bank repository of the NASN.) One of the determining factors for national air surveillance is cost. Some 360,000 filters purchased at an approximate cost of 50¢ per filter were furnished for air monitoring in calendar year 1978. For metal analyses, my estimate is that the overall cost per quarterly composite sample, excluding analysis and data reporting (which of course includes the individual sample cost as well as the cost of compositing strips from these individual samples), is in the vicinity of $200 per sample. This means approximately $1+ million per year is expended in obtaining samples for the NASN, and probably that much more is expended overall in the analysis of these samples and in the reporting of data.

Nature of Particulate Matter. In an endeavor to determine the analytical interferences, massive (i.e., 1 gram) integrated samples were taken by the continuous operation of a hi-volume sampler for a 10-day period in Cincinnati, Denver, St. Louis, Washington, Chicago, and Philadelphia during December 1969 through April 1970. Collections were made using glass-fiber filters, membrane filters, and silver filters to afford total elemental composition. These sampling efforts were conducted by EMSL at all six sites. The analytical techniques used were those obtainable from commercial sources by contract (*18*). The method of choice for a given element was that adjudged to give the most accurate determination considering the standards available and the technique limitations. The overall ambient ranges obtained from these sampling efforts at the six sites are given in Table II for selected elements. For vanadium, one can note that the range in concentrational levels are a factor of 300 while one can note for chlorine, a nonmetal and thus not necessarily germane to the consideration at hand but of interest, a range of some 4000 in levels. Concentrations then of metals can be expected to vary by some 10^4. This is a realistic value when one considers that the 10-day integrated samples dampen the concentrational variances with respect to time, and that only six sites were considered in this short study. (Note also that the sampling efficiency is assumed to be 100%.) The massive samples were analyzed for all the elements except the noble elements and the elements that have no stable nuclides; blank values, where detectable, were compensated for when computing the elemental composition. The levels found are given in Table III. Note that for the elements In, Ru, Rh, Pd, Os, Ir, Pt, and Au, the concentrations in the samples examined were below the limits of

Table II.   Overall Ambient Ranges of Particulates (ng/m^3)

Element	Low	High	Typical
Be	0.01	0.6	0.2
C	28,000	110,000	50,000
N	2,400	9,700	4,000
Si	4,000	45,000	10,000
K	600	10,000	1,000
V	1	300	30
Ni	2	100	20
Fe	1,500	9,000	4,000
As	2	100	4,000
Sn	6	200	50
Pb	300	5,000	200
S	3,000	11,000	5,000
Cl	< 4	13,000	1,000
Br	10	200	100

detection for the methods used which included SS/MS, OES, AAS, XRF, and NAA (*18*). In terms of average values, the data found for the elements routinely analyzed by the NASN are compatible with those given in this study. In Table III, the values given can be taken as representative for that time for metropolitan air of the U.S. cities sampled. Elemental data for the 1970 NASN survey are given in Table VI of Ref. *13*.

In the main, a range of only 100 is found for most elemental concentrations in composite urban samples. However, the analysis of particulate matter for trace elements is much more demanding than the analysis of natural products wherein the range of trace elements in a particular type of specimen is usually comfortably within a factor of 10. One also notes that instead of some 25 or more trace metals being determined, as is the case with particulate matter, an analysis of natural products is usually restricted to 8 or 10 elements. There also is a problem of instrumental or technique bias to contend with in the analyses of particulate matter because of the nature of the complex matrix which results from the number of constituent elements and the range of individual concentrations displayed. This problem is usually not as complex in analyses of samples from other sources. The lack of standards of a nature closely comparable to particulate matter (Standard Reference Materials) poses problems of defining precision and accuracy for elemental analyses as well as for instrumental technique comparisons. Intensive short-term monitoring efforts and long-term monitoring efforts at single sites will not yield information of the nature required to meet long-term assessment objectives. The temporal and site variances are masked in the NASN endeavor because quarterly composites are analyzed. The data therefore are smoothed and smoothed further when reported for trend analysis purposes (*19*). Because of the low levels of some constituents (below the level of detection with many instruments) the method of statistical evaluations of the data should be examined closely. (It is a common practice in computing averages to use 0 when below detectable is reported for a given elemental level, which can give an average value lower than is probable.) It is "in-house" practice to use one-half of the lower limit of detection rather than 0.

Filter Preparation. Currently, for the sake of minimal NASN sample usage, a quarterly composite is constructed from strips from individual filters. A minimum of five samples and not more than seven must be taken per quarter. No two samples shall be taken within the same calendar week. An interval of at least two days must exist between samples. There should be at least one sample and not more than three per month. A composite is oxidized in a low-temperature asher, extracted with mixed (1:3 constant boiling hydrochloric and nitric) acids, the acid extract concentrated and the supernatant taken for analysis. (In some cases

Table III.   Range and Typical Values for Elements in Particulates Collected from Ambient Air[a] (ng/cm^3)

Element	Low	High	Typical
H total	3,000	11,600	5,000
"inorganic"[b]	1,600	4,700	3,000
He		Not sought	
Li	2	20	4
Be	0.01	0.6	0.2
B	3	30	5
C total	28,000	110,000	50,000
"inorganic"[b]	15,000	50,100	25,000
N total	2,400	9,700	4,000
"inorganic"[b]	2,100	7,800	3,000
O		Not determined	
F	20	900	50
Ne		Not sought	
Na	500	12,000	2,000
Mg	1,000	5,000	2,000
Al	1,000	10,000	3,000
Si	4,000	45,000	10,000
P	50	600	100
S total	3,000	11,000	5,000
"inorganic"[b]	1,700	8,700	4,000
Cl	4	13,000	1,000
Ar		Not sought	
K	600	10,000	1,000
Ca	2,000	20,000	6,000
Sc	1	10	1
Ti	150	600	200
V	1	300	30
Cr	3	50	20
Mn	30	200	100
Fe	1,500	9,000	4,000
Co	3	10	5
Ni	2	100	20
Cu	100	3,000	500
Zn	200	3,000	500
Ga	3	8	5
Ge	0.4	10	2
As	2	100	10
Se	2	20	4
Br	10	200	100
Kr		Not sought	
Rb	10	100	20
Sr	20	200	40
Y	1	5	2
Zr	2	20	4
Nb	0.3	4	0.5

Table III. Continued

Element	Low	High	Typical
Mo	0.5	10	1
Tc	----------------------------	Not sought	----------------------------
Ru	0.01	0.1	0.03
Rh	0.01	0.1	0.03
Pd	0.2	1	0.3
Ag	0.5	20	1
Cd	0.3	20	1
In	0.5	2	1
Sn	6	200	50
Sb	2	60	5
Te	0.1	0.2	0.1
I	0.01	0.1	0.4
Xe	----------------------------	Not sought	----------------------------
Cs	0.3	3	1
Ba	10	500	100
La	2	20	5
Ce	3	20	5
Pr	1	5	2
Nd	1	10	5
Sm	0.4	2	1
Eu	0.1	0.5	0.2
Gd	0.3	1	0.6
Tb	0.05	0.2	0.1
Dy	0.2	1	0.2
Ho	0.03	0.1	0.3
Er	0.3	0.5	0.3
Tm	0.03	0.1	0.06
Yb	0.05	0.3	0.1
Lu	0.02	0.3	0.05
Hf	0.04	0.2	0.05
Ta	0.1	0.3	0.2
W	0.1	2	1
Re	0.01	0.06	0.02
Os	0.02	0.06	0.04
Ir	0.01	0.06	0.02
Pt	0.02	0.1	0.04
Au	0.02	0.1	0.06
Hg	0.02	0.1	0.04
Tl	0.01	0.1	0.02
Pb	300	5,000	2,000
Bi	0.2	2	0.6
Po	----------------------------	Not sought	----------------------------
At	----------------------------	Not sought	----------------------------
Rn	----------------------------	Not sought	----------------------------
Fr	----------------------------	Not sought	----------------------------
Ra	----------------------------	Not sought	----------------------------
Ac	----------------------------	Not sought	----------------------------

Table III.   Continued

Element	Low	High	Typical
Th	0.05	1	0.2
Pa	----------------------------	Not sought	----------------------
U	0.05	2	0.2

Transuranic elements not sought

[a] Chicago, Cincinnati, Denver, Philadelphia, St. Louis, Washington, D. C.; NTIS Document PB-220-401, p. 5.
[b] Residue remaining after C_6H_6 extraction.

neutron activation analysis, spark-source mass spectrometric or x-ray fluorescence is used without sample preparation.) For some elements it is obvious that the low concentration of these elements in the particulate matter and the size of the particulate matter sample taken preclude the analysis of these elements in the average sample by the techniques available.

Observations. It is informative to contrast the concentrations of a particular element at selected sites to see what kind of variation can be expected from site to site. Let us consider manganese concentrations found in air in 1964–1965 in two heavily industrialized areas. In Table IV, results obtained in the Birmingham area on a quarterly basis are displayed. Here the temporal variation and level at a given site are roughly within a factor of two of the average value for the study period. The extremes noted here are within a factor of 14. In the Kanawha Valley area (Table V), with only six sampling sites vs. the 10 which existed in the Birmingham area, one can note variances at a site within a factor of eight of the average value; for the sample set, the extreme values reported for a quarter varied by a factor of more than 180. The study period averages in

Table IV.   Manganese Concentrations, Birmingham Area, 1964–1965 (μg Mn/m^3)

Place	Site	Spring	Summer	Fall	Winter	Study Period Average
			Seasonal Averages			
Bessemer	1	0.21	0.18	0.15	0.11	0.16
Birmingham	3	0.33	0.35	0.35	0.18	0.30
Birmingham	4	0.82	0.50	0.58	0.95	0.71
Birmingham	5	0.45	0.72	0.45	0.20	0.46
Birmingham	7	0.69	0.35	0.12	0.45	0.40
Fairfield	1	0.36	0.21	0.15	0.10	0.20
Irondale	1	0.30	0.33	0.12	0.21	0.24
Mt. Brook	1	0.21	0.25	0.19	0.12	0.19
Tarrant	1	0.39	0.44	0.29	0.46	0.40
Vestavia	1	0.27	0.14	0.09	0.07	0.14
Average	—	0.40	0.35	0.25	0.29	0.32

Table V. Manganese Concentrations, Kanawha Valley Area, 1964–1965
Suspended Particulates, μg Mn/m^3

Sampling Site and Number	Fall 1964	Winter 1964-65	Spring 1965	Summer 1965	Study Period Average
Smithers (5)	11.00	6.50	4.50	3.00	6.25
Cedar Grove (7)	4.20	3.50	1.80	0.32	2.45
Kanawha City (13)	3.30	1.10	0.53	0.27	1.30
South Charleston, East (20)	2.00	0.26	0.23	0.17	0.67
St. Albans (24)	0.43	0.44	0.06	0.06	0.25
Nitro (25)	0.59	0.28	0.10	0.09	0.27
Average	3.6	2.0	1.2	0.65	1.9

Birmingham varied from .14 to .71, a factor of approximately five, while in the Kanawha Valley area the study period averages ranged from .25 to 6.3, which differ by a factor of more than 24.

In Table VI data is presented for the annual average of chromium at selected NASN urban sites. Chromium is an element that anthropogenic activity would be expected to contribute to atmospheric levels for the five-year period shown for the seven cities considered. There is no consistency in the temporal pattern exhibited. In a typical case, the annual average is within ± 50% of the five-year average, except for one year. In the most extreme case, Baltimore, the "outliner" here is within approximately a factor of four of the five-year average, and in general, the annual average which is in poorest agreement with the five-year average is a low value. In Table VII the seasonal variation of chromium at selected sites is examined by taking the sum of concentrations for the first and last quarter of the year and dividing that by the sum of the second and third quarters of the year. (One would assume that the first and last quarters of the year include most of the months when the temperature demands heating of occupied structures.) From Baltimore, the city where

Table VI. Annual Average Chromium Concentrations NASN Urban[a]
(ng/m^3)

Location	1965	1966	1967	1968	1969	Five-Year Average
Baltimore, MD	17	99	75	51	101	69
Jersey City, NJ	31	14	36	26	51	32
Pittsburgh, PA	16	16	27	49	50	32
Charleston, WV	35	10	14	52	35	29
Youngstown, OH	8	31	18	44	30	26
Cincinnati, OH	20	17	15	22	35	22
Newark, NJ	9	15	15	16	24	16

[a] Minimum Detectable Concentration = 6 ng/m^3.

Table VII.   Seasonal Variations of Chromium at NASN Sites (Concentration in ng/m³)

	Avg. Concn. Quarters 1 & 4	Avg. Concn. Quarters 2 & 3	Ratio $\dfrac{1+4}{2+3}$
Baltimore, MD			
1965	20.5	15.5	1.3
1966	155.0	45.0	3.4
1967	58.5	93.5	0.6
1969	101.0	103.0	1.0
Jersey City, NJ			
1965	57.5	45.0	1.3
1966	14.0	14.5	1.0
1967	44.5	29.0	1.5
1969	53.5	50.5	1.1
Pittsburgh, PA			
1965	4.5	26.5	0.17
1966	11.0	23.0	0.48
1967	15.5	40.0	0.39
1968	31.0	69.5	0.45
Charleston, WV			
1965	64.5	7.5	8.6
1966	9.0	10.5	0.86
1967	8.5	20.0	0.43
1968	29.5	41.0	0.72

the most extreme variation in the annual averages from the five-year average was noted, one can note that with the exception of 1967, all of the values noted are above unity and the average is about two. In New Jersey, all of the ratios are unity or above, the average is about 1.2, and the values are within 20% of that which bespeaks little difference between the seasons overall. In Pittsburgh, however, the values ranged from .17 to .48 with the average being .36. In Charleston, the values range from an unusual 8.6 in 1965 to a low of .43 in 1967. These two values vary by a factor of 20, which is inexplicable without some knowledge of local occurrences.

In Table VIII the annual averages of nickel at selected NASN urban sites are presented. Here the extreme case is in East Chicago where the low of 35 of 154 are approximately within a factor of two of the five-year average of 76. All of the five-year averages lie between 60 and 100 ng/m³. At selected non-urban sites where minimum detectable concentrations can be expected (Table IX), we find a seasonal variation to be meaningless in some instances because the minimum detectable values

Table VIII.   Annual Average Nickel Concentrations at NASN Urban Sites (ng/m³)

Location	1965	1966	1967	1968	1969	Five-Year Average
New Haven, CT	85	79	92	84	139	96
Philadelphia, PA	122	36	62	84	96	80
East Chicago, IN	154	35	35	55	103	76
Reading, PA	61	38	38	92	122	70
Jersey City, NJ	69	52	75	69	64	66
Newark, NJ	73	49	77	61	55	63
Providence, RI	41	58	37	64	112	62

[a] Minimum Detectable Concentration = 6 ng/m³.

were exhibited. In this regard, only Jefferson County, New York, in 1966 and Cape Hatteras in 1969 show values of less than unity. The variations encountered here are much less than those noted at the urban sites.

On the basis of its boiling point and its common usage, cadmium might be expected to be an element that man contributes to the air in significant quantities. This is borne out elsewhere (20). If one considers the quarterly and annual averages of cadmium in the air of the five most populated cities for a given year (Table X), one can note that in 1969 in New York City the extremes varied by a factor of almost six while in the other cases the values are amazingly consistent for three of the cities. The variation in values in Philadelphia is within a factor of two. For Chicago and Los Angeles, essentially the same level is reported for each quarter. For Detroit, the variation is within 20% of the average. Overall one might ascribe the constancy of cadmium concentrations to a constant anthropogenic activity. The selected sites where averages are of 0.15 μg/m³ in 1969 (Table XI) one can note that in Helena, Montana, where there is a considerable amout of lead ore processing (with which cadmium is associated), variation between quarters are almost a factor of 20 whereas in heavy industrialized East Chicago, Indiana, the concentra-

Table IX.   Seasonal Variation in Nickel Concentrations at Some NASN Non-Urban Sites[a]

	1965	1966	1967	1969
Cape Hatteras, NC	ρ[b]	3.8	1.8	0.5
Calvert County, VT	ρ	1.6	1.2	2.6
Orange County, VT	ρ	1.2	1.6	1.6
Jefferson County, NY	1.8	0.9	1.0	1.6
Clarion County, PA	ρ	1.2	1.3	2.8

[a] Ratio of quarterly concentrations for the colder six months (Quarters 1 and 4) compared with warmer six months (Quarters 2 and 3).
[b] ρ implies an indefinite ratio caused by a value of θ for the denominator or numerator and denominator. θ implies the minimum detectable concentration or less.

Table X. Quarterly and Annual Average Cadmium Concentrations in Air of the Five Most-Populated U.S. Cities, 1969 ($\mu g/m^3$)

City	Quarterly Average				Annual Average
	1	2	3	4	
New York City, NY	0.017	0.023	0.004	0.011	0.014
Chicago, IL	0.015	0.014	0.015	0.015	0.015
Los Angeles, CA	0.006	0.006	0.006	0.006	0.006
Philadelphia, PA	0.010	0.014	0.020	0.015	0.015
Detroit, MI	0.011	0.015	0.014	0.010	0.012

tional range is within a factor of three. In El Paso, Texas, there is a considerable amount of ore processing as might be gathered in the annual average of .1 $\mu g/m^3$ for cadmium; here one can note a difference between levels of a factor of less than three, however.

The metallic element whose concentration in air is most clearly ascribable to man's activities is lead. The level of lead found varies from about 0.1 $\mu g/m^3$ in sparsely populated areas to about 10 $\mu g/m^3$ in areas where automobile traffic is at a maximum. In a study conducted at Los Angeles, two sites were monitored on either side of a freeway which had approximately 200,000 cars per day as a traffic level (21). One site was downind a preponderance of the time. If the concentrations at the downwind site are divided by the concentrations of the site that was usually upwind, one can see, in Table XII, ratios that vary from 1.03 (where the percentage of favorable wind was 28) to 4.3 (where the favorable wind percentage was 68). In Table XIII are shown the results obtained when a massive respirable particulate sampler was field tested. The collector was built for EPA by Battelle Columbus. Particles greater than 20 μm are eliminated with a cyclone. The first stage collects particles by impaction

Table XI. Selected Quarterly and Annual Average Cadmium Concentrations at NASN Sites with Annual Average Concentrations Greater Than 0.015 $\mu g/m^3$, 1969

City	Quarterly Average, ($\mu g/m^3$)				Annual Average ($\mu g/m^3$)	Cadmium in Particulate Matter (%)
	1	2	3	4		
Waterbury, CT	0.008	0.029	0.020	0.023	0.020	2.4
E. St. Louis, IL	0.045	0.013	0.016	0.015	0.022	1.8
E. Chicago, IN	0.017	0.046	0.027	0.024	0.028	1.5
St. Louis, MO	0.013	0.060	0.041	0.031	0.036	1.7
Helena, MT	0.077	0.004	0.005	0.026	0.028	4.8
Newark, NJ	0.015	0.014	0.024	0.099	0.038	5.1
El Paso, TX	0.083	0.150	0.057	0.130	0.105	7.0

Table XII. Monthly Average Lead Concentrations for 1975

Site A $(\mu g/m^3)$	Site C $(\mu g/m^3)$	Difference $C - A$	Ratio C/A	% Favorable Wind Direction
6.0	6.2	0.2	1.03	28.2
4.1	4.8	0.5	1.18	35.2
3.2	5.2	2.0	1.63	43.3
2.8	6.0	3.3	2.14	47.6
1.8	7.4	5.6	4.06	65.6
1.4	8.1	6.8	5.68	71.2
1.9	8.0	6.1	4.27	68.0
2.8	8.4	5.7	3.00	61.6
3.7	8.5	5.0	2.30	59.5
4.2	7.3	3.3	1.74	49.4
4.8	6.6	1.7	1.37	31.6

in a cut centered at 3.5 μm on a slitted plate; a second impaction plate collects particles in a cut centered at 1.6 μm. The "fine" particulate matter is collected by an electrostatic precipitator. All collection surfaces (including the plates of the electrostatic precipitator) are Teflon coated to minimize contamination. The particle size distribution of the fractions collected together approximate the size distribution of the ACGIH respirable curve. In both instances, the least amount of lead was associated with the larger particle size fracton. The particle size distribution is no doubt atypical of that to be expected in ambient air at a site removed from a freeway since these sampling sites are immediately adjacent to the heavily traveled Los Angeles freeway.

It is of interest to contrast the air levels expected from geochemical levels with those found experimentally. In the case of platium, a considerable interest was displayed over the possible contamination of the atmosphere by platinum from exhausts of catalytic converter-equipped automobiles. Since platinum compounds are known to be toxic and could conceivably escape from a catalytic converter into the air, it is possible

Table XIII. Analyses of Collected Particulates from Los Angeles Background Catalyst Study Site
(%)

Sample Designation		Determination
	Site A	Pb
Upwind	1st stage	0.86
	2nd stage	1.13
	Electrostatic	0.93
	Site C	Pb
Downwind	1st stage	2.43
	2nd stage	3.16
	Electrostatic	3.74

that airborne platinum might be a source of environmental problems. (It is of interest that no evidence is known to me to suggest that: (a) platinum is detectable in exhaust effluents or (b) platinum compounds are in fact formed in catalytic converters.) To estimate the probable levels, the geochemical data taken from the literature were examined, and an estimate of platinum in air made of a predicted atmospheric loading, assuming that geochemical levels would prevail in the air, i.e., assuming that the particulate matter has essentially the same platinum level as does the earth's surface. Of course, the samples which were to be investigated did not come from precisely the same spot as the geochemical sample that was taken for estimation purposes, and in some instances the geochemical levels are at best semi-quantitative. A particulate sample of ambient air was investigated by SS/MS by the ACB, and an estimate was made on a semi-quantitative basis. This estimate was expected to be no better than within a factor of 10 of the "true" value.

This comparison expected with the "true" results was based on the precision of the work obtained in standards addition techniques; results are given in Table XIV, indicating 4×10^{-11} g/m³ of Pt of air were obtained. A massive sample was obtained from a bag house on the roof of a campus building close to the Los Angeles freeway in the vicinity of the sampling site. Using the mass of the sample and an average particulate loading for Los Angeles, an air volume was assumed which was used in computing a platinum level in the sample when analytical results were obtained. Both OES and XRF proved to be not sensitive enough for this determination to generate data of any value. AAS gave values less than 3×10^{-11} g/m³ where the geochemical level led something in the 10^{-12} g/m³ range to be expected. Neutron activation analysis is a little more

Table XIV. Environmental Platinum[a]

Sample	Method	Pt Level (g/m³)
Suspended particulate method	SS/MS	4×10^{-11}
Geochemical[b]	Literature	2×10^{-12}
Massive sample[c]	IDMS[d]	2.5×10^{-12}
	NAA[d,e,f]	$\sim 9 \times 10^{-11}$
	AAS[e,g]	$< 3 \times 10^{-11}$
	XRF[d,e]	Below detectable
	OES[d,e,g]	Below detectable

[a] Taken from the Los Angeles Catalyst Study (Reference 20), R. J. Thompson, p. 49.

[b] An average Los Angeles particulate loading was assumed to contain platinum at the average geochemical level given in the literature.

[c] An air volume was assumed from the average Los Angeles particulate loading and the mass of the sample.

[d] Oak Ridge National Laboratory. Isotopic Dilution Mass Spectrometry.

[e] ABC/EMSL/EPA.

[f] Battelle Northwest.

[g] Stewart Laboratories, Knoxville, Tennessee.

sensitive but still had a minimum detectable at the 10^{-11} g/m^3 ballpark. Only isotope-dilution mass spectrometry gave data believed to be usable. (The data given were stated to be sufficiently above the limit of detection to constitute a measurement (22).) The results obtained (2.5×10^{-12} g/m^3) were in excellent agreement with a value predicted from the geochemical literature of 2×10^{-12}. This agreement is no doubt fortuitous but of interest. The primary reason why this point is brought up is to illustrate that, for many elements, to estimate the content of a trace metal, a massive sample and a research endeavor might be required. Monitoring of many trace metals is thus completely impractical if one considers that the monitoring of some 30+ trace metals, as it is conducted on a limited basis at 270 sample sites, requires in the vicinity of $2 million a year for support from certain federal, state, and local agencies.

In summary, note that there are many environmental numbers but few environmental data for trace metals. The only environmental measurements to be accepted as being definitive are those which are made with techniques that have been corroborated by phenomenologically different alternative methods of analysis. When considering trace element distributions, one should think in terms of temporal changes and spatial changes being possible in ranges up to 10^4 and that long-term trend assessment data will consist of either data from samples integrated over long time periods or short-term (24-hour or less) sample data which are smoothed considerably in the averaging process. One should be extremely reluctant to accept as typical data that are obtained from an isolated sample taken over a short-term period or to consider as typical for a large area, data from samples taken over an extended period at one site.

Literature Cited

1. McCarthy, J. H., et al., U. S. Geological Survey, Prof. Paper (1970) **713**, 37.
2. Williston, S. H., *J. Geophys. Res.* (1968) **73**, 7051.
3. Mitchell, W. J., Midgett, M. R., *JAPCA* (1976) **26**, 674.
4. Baldeck, C., Kalb, G. W., Crist, H. L., EPA Report **EPA-R2-73-153**, p. 3, (1973).
5. Spittler, T. M., Thompson, R. J., Scott, D. R., Division of Water, Air and Waste, 165 National Meeting, ACS, April 1973.
6. Long, S. J., Scott, D. R., Thompson, R. J., *Anal. Chem.* (1973) **45**, 2227.
7. Purdue, L. J., Enrione, R. E., Thompson, R. J., Bonfield, B. A., *Anal. Chem.* (1973) **45**, 527.
8. "National Ambient Air Quality Standards," *Fed. Regist.* (1971) **36**, 8186.
9. "Air Pollution Measurements of the National Air Sampling Network 1953–1957," Public Health Service Publication No. 637, 1958.
10. Wedding, J. B., McFarland, A. R., Cermak, J. E., *Environ. Sci. Technol.* (1977) **11**, 387.
11. Clements, H. A., McMullen, T. B., Thompson, R. J., Akland, G. G., *JAPCA* (1972) **22**, 955.
12. Skogerboe, R. K., Am. Soc. Testing Materials, Special Technical Publication 555 (1974) 125.

13. Scott, D. R., Loseke, W. A., Holboke, L. E., Thompson, R. J., *Appl. Spectrosc.* (1976) **30**, 392.
14. Hwang, J. Y., *Anal. Chem.* (1972) **44**, 20A.
15. Thompson, R. J., Morgan, G. B., Purdue, L. J., *At. Absorpt. Newsl.* (1970) **9**, 53.
16. Analytical Chemistry Branch, Environmental Monitoring and Support Laboratory, E.P.A., Research Triangle Park, N.C., unpublished data.
17. Scott, D. R., et al., *Environ. Sci. Technol.* (1976) **10**, 877.
18. Henry, W. M., Blosser, E. R., "Study of the Nature of the Chemical Characteristics of Particulates Collected from Ambient Air," NTIS Document **PB-220 401**, EPA Contract **CPA 22-69-153**, 1970. Battelle Memorial Institute, 1970.
19. "National Trends in Trace Metals in Ambient Air 1965–1974," Publication No. **EPA-450/1-77-003**, 1977.
20. McMullen, T. B., Faoro, R. B., *JAPCA* (1977) 1164.
21. Burmann, F. J., "The Los Angeles Catalyst Study Symposium," Publication No. **EPA-600/4-77-034**, p. 10, 1977.
22. Carter, J., Oak Ridge National Laboratories, private communication.
23. Environmental Monitoring and Support Laboratory, E.P.A., Research Triangle Park, N.C., unpublished data.

RECEIVED January 26, 1978.

Preliminary Report on Nickel-Induced Transformation in Tissue Culture

MAX COSTA[1]

Institute of Materials Science, University of Connecticut, U-136, Storrs, CT 06268

Department of Laboratory Medicine, University of Connecticut Health Center, Farmington, CT 06032

Exposure of secondary cultures of Syrian hamster fetal cells to crystalline Ni_3S_2, Ni_3Se_2, and αNiS induced a concentration-dependent incidence of transformation. Similar exposure to amorphous NiS produced no significant transformation. The Ni_3S_2-transformed colonies were cloned and derived into cell lines. The Ni_3S_2-transformed cell lines differed from normal cells in that: (1) they were immortal in culture (survived over 80 passages while the normal cells survived only 10–15 passages); (2) proliferated in soft agar (0.5% W/V); (3) reached 2–3 times higher cell density at confluency compared with normal cells; (4) had a higher plating efficiency; and (5) doubled at a faster rate compared with the corresponding normal cells. These results indicate that specific nickel compounds induce different incidences of transformation in tissue culture.

A number of metals (cadmium, chromium, nickel, and cobalt) and related compounds have been implicated as carcinogens based upon epidemiological and experimental evidence. These metals are thought to act as primary carcinogens since no metabolic activation appears to be involved. A characteristic of a primary metal carcinogen is that cancers are produced at the point of application. The carcinogenesis of nickel

---

[1] Current address: Department of Medical Pharmacology and Toxicology, College of Medicine, Texas A & M University, College Station, TX 77843.

and related compounds in experimental animals has been well studied using a variety of species and different routes of administration. For example, nickel dust inhalation by guinea pigs results in lung cancers (1). Intraosseous and intrapleural administration of nickel dust to rats and rabbits results in sarcomas (2,3) while intramuscular injections of nickel dust induces rhabdomyosarcomas (4,5). Many other routes of nickel administration have been attempted, resulting in various forms of malignant neoplasms in a variety of animal species.

The carcinogenicity of nickel compounds in industrial workers has been comprehensively reviewed in several publications (6,7,8,9,10). Increased incidences of cancer of the lung and nasal cavities have been documented by epidemiological investigations of nickel-refinery workers in Wales (11,12), Canada (13,14), Norway (15), and Russia (16). The identity of the nickel compounds that induce cancers in nickel-refinery workers remains uncertain although principal attention has been focused upon (a) insoluble dusts of nickel subsulfide (Ni_3S_2) and nickel oxides (NiO; Ni_2O_3), (b) the vapor of nickel carbonyl ($Ni(CO)_4$) and, (c) soluble aerosols of nickel sulfate, nitrate, or chloride (8,10,17).

It is particularly interesting that certain nickel compounds have been found to possess different carcinogenic potencies in experimental animals. For example, the intramuscular administration of crystalline Ni_3S_2 results in 90–100% incidence of rhabdomyosarcomas while rats that received similar doses of amorphous NiS did not develop muscle tumors (18,19).

In Vitro Carcinogenesis Screening

A number of in vitro carcinogenic tests for organic compounds have been developed. A recent evaluation of these various test systems revealed that two of them were highly reliable ($> 90\%$) in predicting whether a compound was carcinogenic or not (20). These two systems were the Ames test and the cell transformation assay (20–39). The Ames test uses bacterial systems to measure the mutagenicity of a particular compound (21,22). This system has been successfully applied to organic secondary carcinogens, activated in vitro by incubation with hepatic microsomes. However, attempts to use the Ames test for metal carcinogenesis testing has yielded success only with a few metal oncogens (23–28). A recent report has described the technical difficulty of using the Ames test to screen for potentially carcinogenic compounds (40). Therefore, attention should be focused on the possible use of tissue culture as a test system for measuring the carcinogenic potency of metal compounds. Two distinctly different types of tissue culture systems are available for the metal cell transformation assay (29–39). One involves the use of a tissue culture cell line, while the second uses a primary cell culture. Use of a cell line has the disadvantage that by definition it is a transformed cell since it is immortal while in culture. However, this is only one of the transformed properties and it may lack many of the other morphological and growth properties that characterize transformed cells. Acquisition of other transformed properties have occurred in cell lines following exposure to carcinogens (29, 34). Examples of this are the transformed cell lines that grow to a fixed density (34). Following periods of exposure to a carcinogen, the cell line can lose this particular property and acquire the noncontact-inhibited property of a transformed cell. Another example is the ordered–disordered growth of cells. Normal cells in general have a very ordered pattern of growth while transformed cells grow in a disorganized fashion. Certain cell lines such as baby hamster kidney-21 (BHK-21) cells have a very ordered pattern of growth, but following exposure to a carcinogen, a certain population of these cells will initiate a disordered pattern of growth (41). Recently Fradkin et al. (29) have exposed BHK-21 cells to $CaCrO_4$ and found that following this treatment some of the cells acquired a disordered growth pattern. The ultimate test of transformation, however, is the ability of cells to produce tumors when they are administered to appropriate hosts. Generally, cell lines which have an ordered pattern of growth are not tumorigenic in vivo. These cells may be rendered tumorigenic and acquire a disordered growth pattern following exposure to a carcinogen. However, morphological transformation in tissue culture is not necessarily the same as carcinogenesis, and rigorous criteria are required to prove that transformation induced in tissue culture represents neoplastic transformation. In fact, a recent study has shown that the various criteria which have been applied to transformation such as disordered growth, plant lectin agglutination, and growth in soft agar does not necessarily indicate neoplastic transformation (43). However, the various properties applied to the evaluation of transformation are only observed following exposure to carcinogens but usually not found when cells are treated with noncarcinogens. Therefore, it must be assumed that the carcinogenesis process in tissue culture represents a complex array of changes of which transformation is an integral part.

The primary culture is the second system used for carcinogenic tests (30–37). It has numerous advantages over the cell line because during the early periods of culture it lacks the transformed properties shown in Table I. Embryonic cells are usually used because they proliferate in tissue culture while cells isolated from older animals are nonproliferative unless they are derived from tumors (30–37). The criteria that apply to cell transformation of these primary cultures are shown in Table I. Many other properties can be used to distinguish a normal cell from a transformed cell, such as chromosome composition, enzyme and protein markers, etc., but the criteria listed in Table I are most

Table I. Common Characteristics of Normal and Transformed Cells in Tissue Culture

	Normal Cell	Transformed Cell
Morphology of cell colony	Not piled up	Piled up
	Contact inhibited	Overgrowth of monolayer
	Ordered growth, no criss-crossing	Disordered growth criss-crossing not present in control
Growth conditions	Does not grow in soft agar	Growth in soft agar
	Definite life span in tissue culture	Immortal
	No tumor proliferation upon administration to appropriate host	Tumor proliferation

commonly used. It is beyond the scope of this chapter to review all the primary tissue culture systems that have been used or applied to transformation testing of organic carcinogens (30–37). There are relatively few reports of metal-induced transformation in tissue culture (29, 30) and this new area is focused upon and investigated in the present study.

Rationale for Using a Tissue Culture System in Metal Carcinogenesis Testing

The use of animals to test the carcinogenic potential of compounds is without question the most reliable and accurate test system available. However, tests using this system are hampered by the long period of time it takes to obtain results (usually two years with rats and longer with other animals such as primates) and the extreme costs involved (one needs to use a large number of animals for every compound and one has to care for these animals over a long time period). Consequently, it is not possible to test all the metal compounds in animal systems. Thus, rapid, inexpensive carcinogenic test systems are extremely important tools to use, not in place of animal testing but in conjunction with it as a prescreening procedure. Having a reliable, inexpensive, and rapid carcinogenic test system can eliminate a great many compounds from animal testing. However, morphological transformation in tissue culture does not directly represent neoplastic transformation. Positive results with this assay should only indicate that the compound might cause cancer until it is tested in animals or until the morphologically transformed colony is cloned and produces tumors in nude mice.

Methods

General Tissue Culture Techniques. Primary cultures of Syrian hamster cells were isolated from embryos after 12–14 days of gestation (in a few cases the embryos were 10–11 days old but for most of the experiments described here, the embryos were 12–14 days old). No differences were found in terms of growth of the cells in culture or transformation when cells were isolated from embryos at these various stages. We used primary Syrian hamster cells in our studies since previous reports have indicated that Syrian hamster fetal cells are less prone to spontaneous transformation while in culture compared with other species such as mouse, rat, etc. (38, 39). The pregnant hamsters were sacrificed by exposure to ether vapors. The hamster was then dipped in a beaker containing a 2% Orsyl disinfectant solution for about 2 min. The hamster was taken out of this solution and the fur was drained of excess disinfectant. All subsequent procedures were performed in a sterile area using sterile techniques. The embryonic sacs were removed and placed in separate 100 by 20 mm Petri dishes which contained Eagle's MEM (Eagles Minimal Essential Tissue Culture Medium obtained from Gibco, Inc.) without fetal bovine serum. The embryos were removed from the sac and placed in another tissue culture dish (100 by 20 mm) containing 10 mL of a trypsin solution (0.5% W/V trypsin 1:250 procine parvovirus tested, Gibco, Inc. in Puck's saline A). The embryos were minced in the trypsin solution using a combination of forceps and scissors. The minced embryos were vigorously pipetted in the trypsin solution to aid in the dispersion of the cells. The trypsin treatment ranged from 15–30 min. Following this treatment the cells were seeded into bicarbonate-buffered Eagle's MEM containing 10% fetal bovine serum (Gibco, Inc. heat inactivated viral screened), 100 units/mL of penicillin, 0.25 μg/mL fungizone, and 100 μg/mL of streptomycin (Gibco, Inc.). Cultures were checked routinely for mycoplasma contamination by the method of Peden (43). In some experiments the cells were seeded into Dulbecco's MEM tissue-culture-medium supplemental with 10% fetal bovine serum (Hy-Clone, Sterile Systems, Inc.). The use of Dulbecco's medium resulted in a slightly higher plating efficiency of the cultures and a slightly better incidence of transformation following exposure to Ni_3S_2. The Dulbecco's medium was used to evaluate the incidence of transformed foci and also for cell cloning and maintenance of the Ni_3S_2-derived cell lines. Monolayer cultures were maintained at 37°C in a 5% CO_2/95% air atmosphere. All cultures were fed two times each week after exposure to the carcinogens.

Metal Transformation Assay. The insoluble metal compounds were weighed, placed in a glass centrifuge tube, sterilized by heating the tubes in an oven at 110°C for 30 min, and the sterilized compounds were suspended in complete tissue media at a concentration of 1 mg/mL. In some experiments (where indicated in the tables) the insoluble metal compounds were sterilized by washing with acetone. Each test-compound suspension was serially diluted from a concentrated mixture in complete media, and the final volume added to the plates was usually > 0.1 mL. Any water-soluble metal compounds used were prepared in complete media and sterilized by passage through millipore filter units (Millex, 0.45 μm) with the aid of a syringe.

Secondary cultures of Syrian hamster cells were plated into 100 by 20 mm plates containing 10 mL of complete medium at a concentration of about 1×10^5 cells per plate. These cells were exposed to the metal compounds at various concentrations and for periods ranging from 4 to 8 days as indicated in the tables. Following this exposure the metal compounds were removed from contact with the cells by washing the monolayers three times with warm Puck's saline A, and the cells were subsequently dislodged from the monolayer by trypsinization. The trypsin acting on the cells was neutralized with complete media and the cells were counted with an electronic particle counter (Coulter Electronics, Inc.). Random verification of cell counts obtained with the Coulter counter were made using a hemocytometer. Five thousand to ten thousand cells for the colony assay or 10^5 cells for the foci assay that had been treated with the various compounds were plated into 35-mm multiwell tissue-culture dishes (Costar, six wells per dish) containing a final media volume of 3 mL. The plated cells contained a number of giant nonproliferative cells which probably served as a feeder culture to permit the growth of cell colonies. This could account for the low plating efficiency and high cell numbers required to obtain 50–100 colonies per plate. No conditioned media was necessary for the formation of colonies. The cultures were fed two times each week with fresh complete media and incubated for a period of two weeks. In some cases the cells were incubated for less than two weeks (10–14 days) depending on the time necessary for cell colonies to form. All cultures were routinely checked every 4–5 days

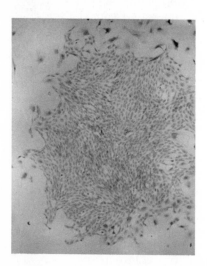 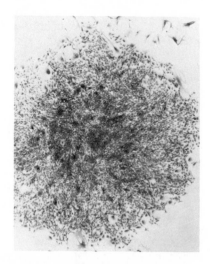

Figure 1. Lightly packed (left) and densely packed (right) normal Syrian hamster fetal cells. Cultures of Syrian hamster fetal cells were obtained from minced whole embryos by trypsin treatment (see Methods for details). These cultures were untreated in parallel with cultures exposed to various carcinogens. The cultures were washed and the cells replated to form colonies in fresh medium. Following two weeks of incubation the cultures were fixed and stained with a crystal violet solution. Note the orderly growth of normal cells in the two colonies.

for the appearance of cell colonies. Following the appropriate incubation time, the plates containing colonies were washed two times with a normal saline solution, fixed with 95% ethanol, and stained with a crystal violet solution (0.5 g crystal violet per liter of ethanol). The morphological appearance of the colonies or foci was examined microscopically. The criterium for transformation used was random cell growth with cell overlapping.

Preparation and Characterization of the Metal Compounds. The Ni_3S_2 was prepared from carbonyl-derived nickel powder and ultrapure elemental sulfur as previously described (*44*). Amorphous NiS powder was produced by ammonium sulfide precipitation from a $NiCl_2$ solution that had been prepared from carbonyl-derived nickel powder and ultrapure HCl. Crystalline Ni_3Se_2 and crystalline αNiS were prepared by mixing stoichiometric amounts of the starting elements, sealing them in evacuated quartz samples, and reacting them at high temperature and pressure. All samples were ground to medium particle size $< 2 \mu m$ in an agate mortar. The crystalline structure of the compounds were determined by powder x-ray diffraction and chemical purity checked by emission spectrographic analysis.

Cloning of Transformed Colonies. In some cases transformed colonies induced by treatment with Ni_3S_2 were isolated so that cell lines could be derived from these transformed cells. The plates containing

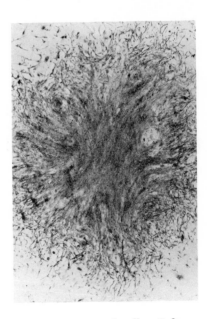

Figure 2. Ni_3S_2-transformed colonies of Syrian hamster fetal cells. Cultures of Syrian hamster fetal cells were exposed two times for two days to Ni_3S_2 (1 μg/ml). The cells were then plated to form colonies for two weeks in fresh medium supplemented with 10% fetal bovine serum. The colonies were fixed and stained with a crystal violet solution and two examples of transformed colonies are shown.

normal and transformed colonies were washed extensively with Puck's Saline A. A transformed colony was selected that was distant from other colonies. A sterile rubber cylinder was placed over the colony and a few drops of trypsin solution was pipetted into the cylinder. The trypsin was incubated on the colony for about five minutes, and then the dislodged cells were removed with a pipette and placed in a new tissue-culture flask with fresh complete medium (Dulbecco's MEM supplemented with 10% fetal bovine serum (Hy-Clone, Sterile Systems, Inc.)). Five cell lines were derived from five Ni_3S_2-transformed colonies. The normal and transformed cells were tested for their ability to proliferate in soft agar medium (0.5%) as previously described (45).

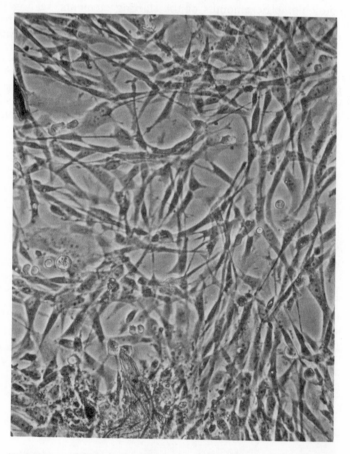

Figure 3.   Cell line derived from a Ni_3S_2-transformed colony. A Ni_3S_2-transformed colony was cloned and derived into a cell line. The figure shows the disordered cell growth of the Ni_3S_2-derived cell lines. This cell line has survived over 80 passages with the same disordered growth and also displays other properties of a transformed culture (see Results).

Results

Morphology of Normal and Transformed Colonies. Figure 1a shows a less tightly packed normal cell colony while Figure 1b shows a denser normal colony. Note that the cells grow in a very orderly fashion in these normal colonies. Two Ni_3S_2-transformed colonies are shown in Figures 2a and 2b. Note the disordered growth of the cells at the edges of the colony. Also note that the cells are densely packed with disordered pat-

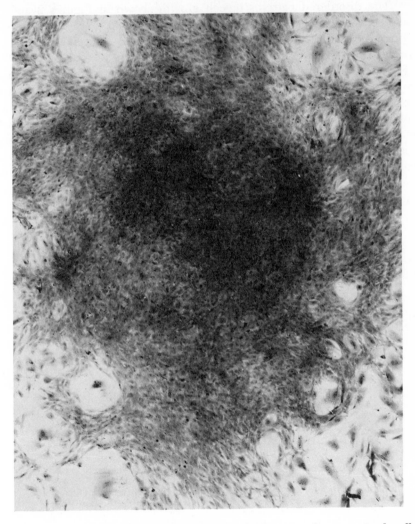

Figure 4. A tightly packed, piled-up normal Syrian hamster fetal cell colony. Treatment with some of the metal carcinogens induced a higher incidence of these tightly packed/piled up normal colonies. This type of colony was also present in untreated cultures.

terns of growth in the center of the colony. Figure 3 shows the morphological appearance of a cell line cloned from a Ni_3S_2-transformed colony. Note the disordered cell growth and cell overlapping found in this Ni_3S_2-transformed cell line. Figure 4 shows a tightly packed cell colony which was present to some extent in the controls but to a greater extent in carcinogen-treated cultures. The colony shown in Figure 4 is normal since there is no disordered growth with cell overlapping as shown in Figures 2 and 3.

The disordered patterns of cell growth shown in Figures 2 and 3 have been uniquely attributed to transformation by a number of investigators (36, 37, 38, 39) and especially DiPaolo and co-workers (36, 37) for Syrian hamster fetal cells. Cell piling or tightly packed cell growth (Figure 4) also has been associated with changes produced by treating cultures with carcinogen, but this growth property is present to some extent in untreated control colonies. We have therefore evaluated our colonies for transformation using cell overlapping and piling as the morphological criteria (Figures 2 and 3) since this is the growth property that has been attributed to transformation of primary Syrian hamster cells (36, 37). However, we have also scored the cultures for cell piling independent of disordered growth to evaluate the effects of the various metals on this parameter as well.

Table II. Transformation of Syrian Hamster Cells by Crystalline Ni_3S_2 [a]

Conditions	Concentration ($\mu g/mL$)	No. of Plates	Transformed Colonies/ Total Surviving Colonies	Percent Transformation	Transformed Colonies/ Plate
Control	0	13	0/323	0	0
Amorphous NiS	1	8	2/204 [b]	1.0	0.25
	2	8	0/125	0	0
	5	10	1/317	0.3	0.10
Crystalline Ni_3S_2	0.1	8	7/318 [c]	2.2	0.88 ± 0.22
	1.0	13	26/362 [d]	7.2	2.00 ± 0.25
	2.0	8	15/174 [d]	8.6	1.88 ± 0.29
	5.0	7	14/174 [d]	10.9	2.71 ± 0.47
	10.0	3	7/125	5.6	2.3

[a] Secondary cultures of fetal hamster cells were exposed to the compounds shown in the table for a period of 6–8 days. The compounds were then removed by washing the cells extensively with Puck's saline A; cells were trypsinized from the monolayer and replated in fresh complete media and allowed to proliferate for two weeks. The cultures were then stained and the morphologies of the cell colonies were evaluated with a light microscope for criss-crossing and piling up. P values were obtained using the χ^2 method, comparing the treated cultures with untreated controls. Error terms are the SEM.
[b] Not statistically significant.
[c] $P < 0.025$.
[d] $P \ll 0.0005$.

Transformation of Syrian Hamster Cells by Metals. When cultures of cells were exposed to two nickel compounds with different carcinogenic potencies, there were differences in the incidence of transformed colonies that formed (Table II). Treatment with amorphous NiS resulted in a low incidence of transformation while treatment with crystalline Ni_3S_2 induced the appearance of numerous transformed colonies (Table II). In this series of experiments, control cultures had no transformed colonies; however, in some experiments there was an occasional incidence of spontaneous transformation. The induction of transformation by

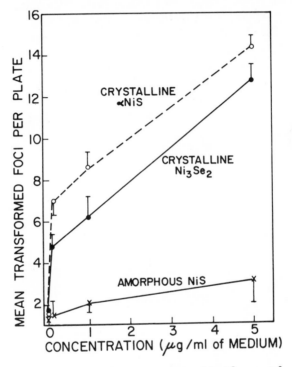

Figure 5.   Induction of transformed foci by crystalline αNiS and crystalline Ni_3Se_2.

Cultures of secondary Syrian hamster fetal cells were prepared with 10^5 cells per 35-mm plates. Following a 24-hr period of attachment the monolayer was treated two times for two days with various concentrations of the metal compounds as shown in the figure. The metal compounds were then removed by washing the monolayer extensively with saline A and the cultures were incubated for 18–21 days with fresh Dulbecco's medium supplemented with 10% fetal bovine serum. The cultures were refed two times per week. The plates were then fixed and stained as described in Methods. Acetone was used to sterilize the metal compound. The number of transformed foci was determined per plate. Each point shown in the mean ± SEM for at least six plates.

Ni_3S_2 depended upon concentration from 0.1 to 5 μg/mL (Table II). Figure 5 shows the incidence of transformed foci per plate as a function of treatment with various concentrations of Ni_3Se_2, αNiS, and amorphous NiS. Note the linear relationship in the incidence of transformed foci from 0.1 to 5.0 μg/mL treatments with crystalline Ni_3Se_2 and crystalline αNiS (Figure 5). Again amorphous NiS resulted in a very low incidence of transformed foci. The experiments shown in Figure 5 were carried out in Dulbecco's MEM instead of Eagle's medium. Similar results in the colony assay were obtained in Dulbecco's medium as in Eagle's medium (Table II); however, Dulbecco's medium yield slightly higher incidence of transformation compared with Eagle's medium. Note also the slightly higher incidence of transformation induced by amorphous NiS in the foci assay (Figure 5) compared with the colony assay (Table II). Relative to the other nickel compounds tested, amorphous NiS is a weak inducer of transformation. Preliminary experiments examining a number of other metal compounds for transformation revealed a good correlation between carcinogenic potency and incidences of transformation in our culture

Table III.   Transformation of Syrian Hamster

Compound Added	Concentration (μmol/L)	No. of Plates
Control	0	3
NaCl	10	3
KCl	10	3
$MgCl_2$	10	3
$CaCl_2$	1	3
	10	3
$CrCl_3$	1	3
	10	2
$CaCRO_4$	1	3
$MnCl_2$	1	3
	10	3
$CoCl_2$	1	3
	10	3
$NiCl_2$	1	3
	10	2
$CuSO_4$	1	3
	10	3

[a] Syrian hamster fetal cells were treated with the various metal compounds shown in the table for 6–8 days. The compounds were removed by washing the cells with saline A. The cells were trypsinized and 10,000 cells were replated with fresh media into 35-mm plates and allowed to proliferate for two weeks. The plates containing

system (Table III). Metals such as Cr^{++}, Co^{++},, and Ni^{++} which are carcinogenic and Cu^{++} or Mn^{++} which are mutagenic induced transformed colonies while metals which are not carcinogenic such as Na^+, K^+, or Ca^{++} did not induce transformation (Table III). Further work with these metals using larger samples are necessary to confirm these results.

Nickel subsulfide (Ni_3S_2) also induced considerably more piled-up, tightly packed cell colonies than amorphous NiS-treated cultures or untreated control cultures (Figure 4, Table IV). All of the transformed colonies were piled up but not all of the piled-up, tightly packed cell colonies were transformed. There were also a considerably large percentage of these piled-up, tightly packed cell colonies in untreated controls (Table IV). These results are probably attributable to the mixed cell population present in cell cultures derived from whole embryos. Selection of cells by Ni_3S_2 is unlikely since in repeated experiments Ni_3S_2 at concentrations of 0.1, 1, or 2 $\mu g/mL$ did not affect the plating efficiency but induced transformation at these concentrations. Higher concentrations of Ni_3S_2 (10 $\mu g/mL$) did, however, reduce the plating efficiency.

Cells by Diverse Metal Salts (Preliminary Data) [a]

Transformed Colonies/ Total Colonies	Percent Transformed	Transformed Colonies/Plate
0/66	0	0
0/43	0	0
0/59	0	0
1/76	1.3	0.33
0/65	0	0
0/60	0	0
1/42	2.3	0.33
1/32	3.1	0.50
3/91	3.3	1
1/102	1.0	0.33
1/61	1.6	0.33
1/76	1.3	0.33
3/82	3.7	1.0
2/85	2.3	0.66
2/51	4.0	1.0
1/70	1.4	0.33
1/42	2.3	0.33

the cells were fixed and stained with crystal violet. Colonies were scored for transformation by examinining the cells for criss-crossing/piling up using a light microscope. All plates were evaluated blindly.

Table IV. Percentage of Piled-Up, Tightly Packed Colonies
Induced by Crystalline Ni_3S_2 and Amorphous NiS[a]

Compound Added	Concentration ($\mu g/mL$ of media)	Number of Plates	Piled-Up Tightly Packed Colonies (% of surviving colonies)
Crystalline Ni_3S_2	1[b,c]	13	33 ± 4
	2[d]	8	31 ± 3
	5[e]	7	39 ± 3
	10	3	48 ± 9
Amorphous NiS	1	8	15 ± 3
	2	8	15 ± 4
	5	10	13 ± 3
No Additions	—	14	10 ± 2

[a] Secondary cultures of Syrian hamster cells were exposed to crystalline Ni_3S_2 or amorphous NiS for 6–8 days. The media containing the metals was removed and the cells washed extensively with Puck's Saline A, trypsinized, and replated to form colonies (10,000 cells per 35-mm tissue culture Petri dish) for two weeks. The cultures were fixed, stained, and scored for normal or piled-up, tightly packed colonies. The percentage of piled-up, tightly packed cell colonies per total surviving colonies shown in the table is the mean $\pm$ SEM for five separate experiments. The statistical significance of the differences between Ni_3S_2 vs. untreated or NiS-treated for each respective concentration was tested using the students' t-test. A further and more detailed report of this data will be presented (Costa, M., Nye, J., and Sunderman, F. W., Jr., manuscript in preparation).
[b] Ni_3S_2 differs from untreated, $P < 0.0005$.
[c] Ni_3S_2 differs from NiS, $P < 0.005$.
[d] Ni_3S_2 differs from NiS, $P < 0.005$.
[e] Ni_3S_2 differs from NiS, $P < 0.0005$.

Verification of Ni_3S_2-Induced Transformation. Cell lines were derived from the Ni_3S_2-transformed colonies. The cell lines retained the characteristic disordered growth shown in Figure 3 through a number of passages (50–80). The normal cells, however, usually died out after 10–15 passages and never acquired the transformed morphology. The Ni_3S_2-transformed cell lines formed colonies in soft agar medium, reached 2–3 times higher cell density at confluency compared with normal cells, had a higher plating efficiency, and doubled faster than normal cells. The normal cells could not form colonies in soft agar medium. A photograph of a Ni_3S_2-derived cell line forming a colony in soft agar medium is shown in Figure 6. These various criteria and especially colony formation in soft agar have been used extensively to verify transformation. Recently, we have found that one of the Ni_3S_2-derived cell lines produced a 100% incidence of tumors in nude mice at the site of injection. Similar subcutaneous injection of normal cells did not produce any tumors in nude mice (46). A number of other cell lines derived from Ni_3S_2-transformed

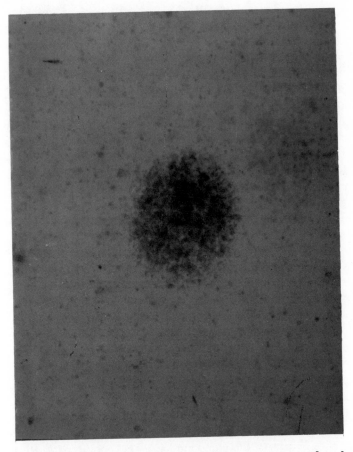

Figure 6. *Colony growing in soft agar from Ni_3S_2-induced cell line. A Ni_3S_2 transformed cell line was plated (10^4–10^6 cells in 60-mm plastic tissue culture dishes) to form colonies in soft agar medium (45). The figure shows the appearance of a colony growing in the soft agar.*

colonies produced tumors in nude mice. Three other cell lines derived from Ni_3S_2-transformed colonies have also produced tumors in nude mice.

Discussion

Carcinogenic metal compounds such as crystalline Ni_3S_2 induced a dose-dependent statistically significant incidence of transformation in secondary cultures of Syrian hamster cells. A chemically related compound, amorphous NiS, which is not carcinogenic, resulted in no statistically significant incidences of transformed colonies. Untreated control cultures also had no statistically significant incidence of transformation. We have verified that the morphologically transformed colonies induced by Ni_3S_2 were in fact "transformed." The Ni_3S_2-transformed colonies gave rise to immortal cell lines which differed in a number of growth properties compared with the normal cell. The normal cells could only be passed 10–15 times while the Ni_3S_2-transformed cells at this writing have survived about 80 passages. The best evidence that Ni_3S_2 induced transformation was that the Ni_3S_2-transformed cell lines proliferated in soft agar medium (Figure 6), and these cell lines produced tumors in nude mice at the site of injection (46). Further observation of these mice followed by autopsy and pathological examinations will have to be performed to evaluate and understand the nature of these tumors (46). We are currently testing tumorous cell lines derived from Ni_3S_2-transformed colonies for their tumorigenicity in nude mice. If a significant number of cell lines derived from Ni_3S_2-transformed colonies produce tumors in nude mice then it can be stated that Ni_3S_2 induced neoplastic transformation in tissue culture. The importance of demonstrating tumor proliferation in animals has recently been emphasized (42). Cell hybrids derived from a neoplastic-transformed cell and a normal cell gave rise to a cell line which had many of the properties of a transformed cell but did not produce tumors in nude mice (42). These results indicate that tumor proliferation in experimental animals is under separate genetic control from transformation, and the only conclusive test for neoplastic transformation is tumor proliferation in experimental animals.

Crystalline Ni_3Se_2 and crystalline αNiS also induced a dose-dependent incidence of transformed foci (Figure 5). It is interesting that amorphous NiS was a weak transforming agent while crystalline αNiS induced a high incidence of transformation. These results suggest that the crystal structure of the compound may be important in determining its ability to induce transformation. Both crystalline Ni_3S_2 and crystalline Ni_3Se_2 are potent carcinogens in experimental animals ((19) and personal communication from F. W. Sunderman, Jr.). Amorphous NiS does not induce tumors in experimental animals (19) and the tumorigenic potency of crystalline αNiS is not known. However, the results of this study would predict that it would be a potent carcinogen.

It is unlikely that the morphological transformation induced by Ni_3S_2, Ni_3Se_2, or crystalline αNiS is attributable to selective toxicity because: (1) it occurred at concentrations of these compounds which did not significantly affect plating efficiency (0.1 or 1 $\mu g/mL$); (2) the transformation increased over a three- to fivefold range from 0.1 to 5 $\mu g/mL$, and none of these compounds increased or decreased plating efficiency more than two- to two and one-half-fold over the same concentration range in a series of about 10 experiments; (3) the morphologically transformed colonies induced by exposure to Ni_3S_2 gave rise to cell lines from which transformation was verified by growth in soft agar and other tests described in results section.

The assay system described in the present study should be useful in screening potential metal carcinogens for their ability to induce transformation. However, further work is required to demonstrate and validate that each of the various metals induce neoplastic transformation in tissue culture. A statistically significant number of transformed colonies induced by exposure of cells to each of the metal compounds must be cloned and derived into cell lines. The tumorigenic potency of these various cell lines must be tested in nude mice.

The transformation of Syrian hamster cells induced by exposure to the metal compounds was very reliable in a series of experiments. However, a large number of plates must be used in order to obtain dose–response relationships and reliable values for a particular compound. In some cases there was spontaneous transformations in the control untreated cultures. However, the incidence of these was so low (1–3 out of 1000 colonies) and not statistically significant so that it would not interfere with the use of this assay for carcinogenesis screening since many of the metal carcinogens induced a statistically significant incidence of transformation that was an order of magnitude higher than any spontaneous transformations. The usefulness of the system described for predicting carcinogenic potency of metal compounds appears promising, and further work to test a number of different metal compounds should validate the systems general applicability to metal carcinogenesis testing.

Acknowledgment

The technical assistance of Julie S. Nye is gratefully acknowledged. I would also like to thank Mary Roche for the typing of this manuscript. I would like to thank E. Kostiner and F. W. Sunderman, Jr. for their consultation on this project. This work was supported by research grant No. ES 01677-01 (Young Environmental Scientist Health Research Grant to Max Costa) from the National Institutes of Environmental Health Sciences.

Literature Cited

1. Hueper, W. C., *Arch. Pathol.* (1958) **65**, 600.
2. Hueper, W. C., *Tex. Rep. Biol. Med.* (1952) **10**, 167.
3. Hueper, W. C., *J. Natl. Cancer Inst.* (1955) **16**, 55.
4. Heath, J. C., Webb, M., *Br. J. Cancer* (1967) **21**, 768.
5. Heath, J. C., Daniel, M. R., *Br. J. Cancer* (1964) **18**, 261.
6. "Nickel and Nickel Compounds," *IARC Mono. Eval. Carcinog. Risk Chem. Man* (1976) **11**, 75.
7. Fishbein, L., *J. Toxicol. Environ. Health* (1976) **2**, 77.
8. Sunderman, F. W., Jr., *Prev. Med.* (1976) **5**, 279.
9. Sunderman, F. W., Jr., "Advances in Modern Toxicology," p. 257, Hemisphere, Washington, DC, 1977.

10. Sunderman, F. W., Jr., Coulston, F., Eichorn, G. L., Fellows, J. P., Mastro-matteo, E., Remo, H. T., Samitz, M. H., *Nickel,* National Academy of Science, Washington, DC, 1975.
11. Doll, R., Mathews, J. D., Morgan, L. G., *Br. J. Ind. Med.,* in press (1977).
12. Doll, R., Morgan, L. G., Speizer, F. E., *Br. J. Cancer* (1970) **24,** 623.
13. Mastromatteo, E., *J. Occup. Med.* (1967) **9,** 127.
14. Sutherland, R. B., Summary report on respiratory cancer mortality at the INCO Port Colborne, Refinery Department of Health, Toronto, Ontario, (1959) 153.
15. Pederson, E., Høgetveit, A. C., Andersen, A., *Int. J. Cancer* (1973) **12,** 32.
16. Saknyn, A. V., Shabynina, N. K., Gig. Tr., *Prof. Zabol. Khim. Prom.* (1973) **17,** 25.
17. Høgetveit, A. C., Barton, R. T., *J. Occup. Med.* (1976) **18,** 805.
18. Gilman, J. P. W., *Cancer Res.* (1962) **22,** 158.
19. Sunderman, F. W., Jr., Maenza, R. M., *Res. Commun. Chem. Pathol. Pharmacol.* (1976) **14,** 319.
20. Purchase, I. F. H., Longstaff, E., Ashby, J., Styles, J. A., Anderson, D., Lefevre, P. A., Westwood, F. K., *Nature (London)* (1976) **264,** 624.
21. Ames, B. M., Sims, P., Grover, P. L., *Science* (1972) **176,** 47.
22. McCann, J., Ames, B. M., *Proc. Natl. Acad. Sci. U.S.A.* (1976) **72,** 5135.
23. Venitt, S., Levy, L. S., *Nature (London)* (1974) **250,** 493.
24. Nishioka, H., *Mutat. Res.* (1975) **31,** 185.
25. Bonatti, S., Meimi, M., Abbondadolo, A., *Mutat. Res.* (1976) **38,** 147.
26. LoFroth, G., Ames, B. H., Abstr. Eighth Meeting of Environ. Mutagen. Soc., Colorado Springs, CO (1977).
27. Tindall, K. R., Warren, G. R., Skaar, P. D., Abstract, 8th Meeting Mutagen Soc., Colorado Springs, CO (1977).
28. Monti-Bragadin, C., Tamaro, M., Bamfi, E., Zassinovich, G., *Proc. Int. Symp. Am. Chem. Chem. Toxicol. Met., Monte Carlo, Monaco,* 1977.
29. Fradkin, A., Janoff, A., Lane, B. P., Kuschner, M., *Cancer Res.* (1975) **35,** 1058.
30. Casto, B. C., Piecynski, W. J., Nelson, R. L., Dipaolo, J. A., *Proc. Am. Assoc., Cancer Res., 1976,* No. 46, 17.
31. Berwald, Y., Sachs, L., *J. Natl. Cancer Inst.* (1965) **35,** 641.
32. Casto, B. C., Janosko, M., DiPaolo, J. A., *Cancer Res.* (1977) **37,** 3508.
33. Chen, T. T., Heidelberger, C., *Int. J. Cancer* (1969) **4,** 166.
34. DiPaolo, J. A., Takano, K., Popescu, N., *Cancer Res.* (1972) **32,** 2686.
35. Reznikoff, C. A., Bertram, J. S., Brankow, D. W., Heidelberger, C., *Cancer Res.* (1973) **33,** 3234.
36. DiPaolo, J. A., Donovan, P. J., Nelson, R. L., *J. Natl. Cancer Inst.* (1969) **42,** 867.
37. DiPaolo, J. A., Donovan, P. J., Nelson, R. L., *Nature (London)* (1971) **236,** 240.
38. Heidelberger, C., Bushell, P. F., *Gann. Monogr. Cancer Res.* (1970) **17,** 39–58.
39. Berwald, Y., Sachs, L., *Nature (London)* (1963) **250,** 1182.
40. Ashby, J., Styles, J. A., *Nature (London)* (1978) **271,** 352.
41. Stoker, M., MacPherson, I., *Nature (London)* (1964) **203,** 1355.
42. Stanbridge, E. J., Wilkinson, J., *Proc. Natl. Acad. Sci. U.S.A.* (1978) **75,** 1466.
43. Peden, K. W. C., *Experentia* (1975) **319,** 1111.
44. Sunderman, F. W., Jr., Kasprzak, K. S., Lau, T. J., Minghetti, P. P., Maenza, R. M., Becker, N., Onkelinx, C., Goldblatt, P. J., *Cancer Res.* (1976) **36,** 1970.
45. MacPherson, I., Montagnier, L., *Virology* (1965) **23,** 971.

RECEIVED November 29, 1977.

Simultaneous Determination of Trace Elements in Urine by Inductively Coupled Plasma–Atomic Emission Spectrometry

W. J. HAAS, JR., V. A. FASSEL, F. GRABAU IV,
R. N. KNISELEY, and W. L. SUTHERLAND

Ames Laboratory, U.S. Department of Energy and Department of Chemistry, Iowa State University, Ames, IA 50011

Inductively coupled plasma-atomic emission spectrometry was investigated for simultaneous multielement determinations in human urine. Emission intensities of constant, added amounts of internal reference elements were used to compensate for variations in nebulization efficiency. Spectral background and stray-light contributions were measured, and their effects were eliminated with a minicomputer-controlled background correction scheme. Analyte concentrations were determined by the method of additions and by reference to analytical calibration curves. Internal reference and background correction techniques provided significant improvements in accuracy. However, with the simple sample preparation procedure that was used, lack of sufficient detecting power prevented quantitative determination of normal levels of many trace elements in urine.

Other authors in this volume have addressed the importance of trace and ultratrace quantities of various substances as they relate to nutritional, environmental, and occupational aspects of human health. This chapter will focus more on the development and application of a specific analytical procedure, based on inductively coupled plasma–atomic emission spectrometry (ICP–AES), for the rapid, simultaneous determination of a number of trace elements in human urine.

0-8412-0416-0/79/33-172-091$05.00/0

Table I.   Trace Element Excretion in Human Urine

Element	Range of Reported Values ($\mu g/24\,hr$)	Model Value ($\mu g/24\,hr$)
Al	20–1000	100
As	0–460	50
B	60–9200	1000
Be	0.24–2.8	1
Cd	2–110	100
Co	0.5–330	200
Cr	0.05–160	70
Cu	0–440	50
Fe	60–640	225
Mn	9–940	30
Ni	1.4–330	11
Pb	1–170	45
Se	0–1390	50
V	0–224	15
Zn	5–1800	500

Inductively coupled plasma–atomic emission spectrometry is an attractive analytical approach for this purpose for a number of reasons (1). First, ICP–AES is readily applicable for the simultaneous determination of a large number of elements. One advantage of this simultaneous multielement determination capability is that it facilitates the measurement of elemental concentration ratios. The determination of these ratios is especially important for the elucidation and understanding of antagonistic and/or synergistic interactions of the elements. Another obvious advantage of the simultaneous analysis capability is that the analysis time and the required amounts of sample are minimized. Second, ICP–AES is attractive because it has exceptional powers of detection. The limits of detection for most elements are on the order of one nanogram per milliliter (1 ppb) when ultrasonic nebulization is used (2). Third, ICP–AES has an unusually wide useful analytical range. When appropriate readout electronics are used, typical analytical calibration curves are useful over concentration ranges which cover four to six orders of magnitude. Thus, the sample dilutions which are so often required to obtain responses which lie within the analytically useful range of atomic absorption spectrometry, for example, are unnecessary for ICP–AES determinations. Fourth, ICP–AES has been shown (3, 4, 5) to be relatively free of the common chemical interelement interferences which have been encountered, for example, in atomic absorption and atomic fluorescence spectroscopy. Finally, the precision and sample analysis throughput rate of ICP–AES are attractive. The precision of analytical results which can be obtained by ICP–AES is comparable with that achieved in atomic absorption spectroscopy and the analysis throughput time for ICP–AES multielement determinations is comparable, for example, with that for single element determinations by flameless atomic absorption.

An important question in the development and application of any new analytical procedure is: what is the state of the art? One aspect of the state of the art with respect to the determination of trace elements in urine is illustrated in Table I. The table shows the range of reported analytical values and the "model values" for 24-hour urinary excretion of a number of trace elements, as compiled from the report of the Task Group on Reference Man (6). The main feature of the information in Table I is that the range of reported "normal" values covers two orders of magnitude for most of the elements listed. There may be considerable variation in the normal excretion of trace elements in urine, but the extent of the ranges shown in the table also may be significantly and artificially enhanced by wide variations in the accuracy of the analytical results that were surveyed. The latter view is supported by the data

Table II.   Comparison of Trace Element Concentrations for Human Urine with ICP–AES Limits of Quantitative Determination

Element	Range of Reported Concentrations[a] (ng/mL)	"Model" Urine Concentration[a] (ng/mL)	Limit of Quantitative Determination[b] (ng/mL)
Al	14–710	70	3
As	0–330	36	15
B	43–6600	710	21
Be	0.17–2	0.7	0.1
Cd	1.4–79	70	0.5
Co	0.36–240	140	10
Cr	0.036–110	50	2
Cu	0–310	36	0.3
Fe	43–460	160	4
Mn	6.4–670	21	0.08
Ni	1–240	8	2
Pb	0.71–120	32	8
Se	0–990	36	8
V	0–160	11	0.7
Zn	3.6–1300	360	8

[a] Concentration calculated for model excretion of 1.4 L/24 hr.
[b] The limit of quantitative determination is defined as the least concentration of the analyte which can be determined with an expected precision of $\pm 10\%$. The values reported here are equal to five times the experimental limit of detection, $c_L(3s_{b1})$, for the elements in 1% nitric-acid solution (9).

presented earlier in Chapter 1 by Mertz (7) for the case of vanadium in peas and for iron, chromium, and zinc in bovine liver standard reference material. Also, similarly wide ranges of results have been reported for the determination of a number of trace elements in a biological reference material when a single analytical method, namely neutron activation analysis, was used (8). Hence, even in the absence of round-robin tests on the analysis of urine, there is a clear need for improved accuracy in the determination of trace elements in urine and in many other biologically important sample materials.

The state of the art with respect to the potential of ICP–AES for the determination of trace elements in urine can be assessed by comparing the "model" concentration and range of concentrations for each of the elements (computed from the data in Table I) to the corresponding experimental limit of quantitative determination (LQD) for that element in aqueous solutions with 1% (vol) nitric acid, as is done in Table II. If the trace element concentrations in urine are as shown in the table, and if the LQDs which can be achieved for urine samples are not substantially different from those observed for 1% nitric acid solutions, then ICP–AES should be applicable for the quantitative determination of many of the trace elements occurring in urine.

Experimental

Plasma. The heart of any ICP–AES system is the plasma itself, a luminous, partially ionized, gaseous discharge. An artist's drawing of an inductively coupled plasma is given in Figure 1, along with a partial view of the necessary plasma torch and induction coil. Detailed information regarding the operation of an analytically useful plasma has been given elsewhere (10, 11). A brief overview of some of the qualities of the inductively coupled plasma which make it a superior analytical source and the conditions under which it was operated for the determination of the trace elements in urine are given below.

The plasma is a dynamic system which furnishes a localized, chemically inert, high-temperature environment which is very effective for the vaporization, atomization, ionization, and excitation of sample materials that are injected into it. Continuous radiofrequency power input is required and continuous flows of argon are necessary for sustaining the plasma, for cooling the walls of the plasma torch, and for introducing sample material. In our case, the sustaining power (1.5 kK) is supplied by a 27.12 MHz radiofrequency generator (Model MN-2500-E, Plasma–Therm, Inc.); the power is coupled into the plasma via a water-cooled induction coil. The plasma resides within the coil but is insulated from it by the outer tube of the quartz plasma torch. The hottest part of the plasma is a toroidal volume within (and extending above) the load coil. The outer tube of the torch is shielded from the hottest region of the plasma by a high-velocity (5 m/sec) flow of argon (16 L/min).

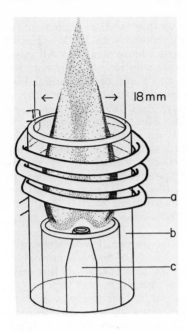

18 mm

Figure 1. Artist's drawing of an inductively coupled plasma. (a) Induction coil, (b) coolant argon, (c) sample aerosol injection tube.

Sample aerosol is transported to the plasma with an argon-gas stream (1 L/min) and is injected into the central channel of the plasma through the aerosol-injection tube of the plasma torch. The sample material is rapidly vaporized, atomized, and excited as it passes through the surrounding, higher-temperature region of the plasma.

Solution Preparation. REAGENTS. Stock solutions for the elements listed in Table I were prepared by dissolving pure metals or reagent-grade oxides in dilute acid. Deionized distilled water (DDW) was prepared with a Barnstead PCS system (Boston, MA). The multielement stock solution and all other working solutions were prepared in a Class 100, horizontal laminar-flow work station (Agnew-Higgins, Inc.), using Class A volumetric glassware. The glassware was leached for at least 24 hours with 1/10 (v/v) HCl/DDW, and rinsed four times with DDW immediately before use.

REFERENCE SOLUTIONS. Three different sets of reference solutions, corresponding to the matrix types shown in Table III, were used. The desired concentrations of internal reference elements (gallium and yttrium) and analyte elements were obtained by adding appropriate,

Table III. Reference Solution Matrices for Urine Analysis

Matrix Label	2% (w/v) NaCl (mL)	Conc HCl (mL)	+DDW to Volume (mL)
DDW	0	1	100
1% NaCl	25	1	100
2% NaCl	50	1	100

small volumes of internal reference solution and multielement stock solution, respectively, before diluting to volume with DDW. The NaCl was ultrapure grade (Ventron Corp.); hydrochloric acid was AR grade.

SAMPLE SOLUTIONS. The urine sample solutions all originated from one composite urine sample. The latter was made by combining 150-mL aliquots of the entire first-morning voids of 16, presumably healthy, male subjects. Aliquots of the composite were used to prepare three sets of sample solutions as indicated in Table IV. The solutions in each set were "spiked" with internal reference elements and with appropriate volumes of multielement stock solution to make a (multiple) standard addition series (12). The final gallium and yttrium internal reference-element concentrations were 1.0 and 0.1 mg/L, respectively, for both the reference and standard addition solutions. The added analyte concentrations for both the reference and standard addition solutions were 0, 2, 4, 10, and 20 times the approximate "normal" analyte concentrations for urine, as listed in Table II.

The important features of the sample preparation procedure were as follows. First, the samples were acidified to dissolve normal urine precipitates and to prevent analyte loss by adsorption on the walls of the sample containers (13). Second, the procedure was kept as simple as possible so that the risk of contamination and/or loss was minimized. Third, dilute, normal, and concentrated series of solutions were used to simulate actual urine samples with a wide range of total dissolved solids. Fourth, because the rate of sample nebulization and the corresponding rate of sample introduction into the plasma can be affected by changes in the amount of total dissolved solids, internal reference elements were included in each sample and reference solution. The use of analyte/internal reference element net intensity ratios provided a means of correcting for possible differences in sample introduction rate according to the internal reference principle (14, 15). Finally, because all of the sample solutions introduced into the plasma were derived from one composite, the different series were known to have trace element concentrations which were related to each other by known dilution factors (see Table IV).

Nebulizer. For the present studies, the sample aerosol was generated with a 3-MHz ultrasonic nebulizer, as illustrated in Figure 2, and the sample aerosol was desolvated prior to its injection into the plasma. A complete description of the nebulizer and desolvation system has been given elsewhere (2). The main virtue of ultrasonic nebulizers is that their efficiency of nebulization (mass of sample injected into the plasma/ mass of sample fed to the nebulizer) is 10 to 12 times greater (2) than that of common pneumatic nebulizers which are suitable for use with

Table IV. Preparation of Urine Sample Solutions

Sample Solution	Urine Composite (mL)	Conc HCl (mL)	+ DDW to Volume (mL)	Dilution Factor
Dilute	20	1	100	5.0
Normal	50	1	100	2.0
Concentrated	90	1	100	1.11

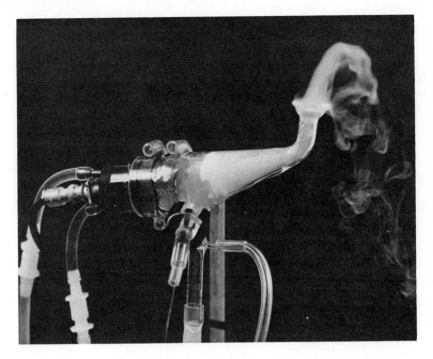

Figure 2. Photograph of the ultrasonic nebulizer in operation

the plasma. Ultrasonic nebulizers, however, also introduce substantially greater amounts of water or other sample solvent into the plasma, as compared with pneumatic nebulizers; the increase in total sample load can produce a significant cooling of the plasma. Consequently, the best detecting powers for ultrasonic nebulization, so far, have not been achieved unless the sample aerosol has been desolvated prior to its injection into the plasma. When desolvation is used, the powers of detection which can be achieved with ultrasonic nebulization are generally an order of magnitude superior to those observed for pneumatic nebulization.

Data Acquisition. The optimum height for observation of the characteristic emission lines of the sample elements is approximately 15 mm above the load coil. At this height the emission spectra of most sample constituents are strongest relative to the continuum emission of the plasma, and the plasma itself is still well shielded from the atmospheric surroundings by its protective sheath of argon gas. In our case, we observe the emission spectra from a 6-mm-high zone of the plasma, centered 15 mm above the load coil. As illustrated in Figure 3, light from this zone is dispersed and detected in a polychromator (Modified Model 127, Applied Research Laboratories). As presently configured (*11*), the polychromator is equipped with 47 fixed exit slits that are precisely positioned to isolate the analytical emission lines of interest and with a corresponding number of RCA 1P28-type photomultiplier

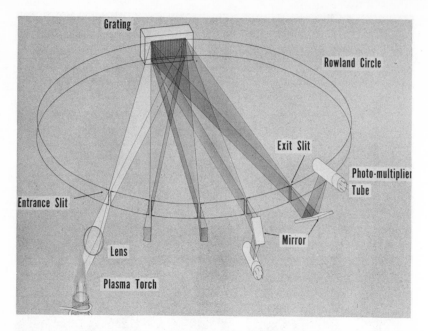

*Figure 3.   Schematic of the inductively coupled plasma—polychromator
arrangement*

detectors. This configuration provides at least one exit slit–detector combination for each of 32 different analyte elements. For simultaneous
multielement determinations, any 20 of the photomultiplier outputs can
be connected to the polychromator readout system through a quick-
connect patchboard arrangement.

The readout system consists of 20 remotely programmable current-
to-voltage (C–V) converters (Model #18000, Keithley Instruments), a
multiplexer-A/D converter (Model 721, Zeltex, Inc.), and a minicomputer (Model PDP-8/e, Digital Equipment Corp.). The voltage signals
from all of the C–V converters are multiplexed into the A/D converter,
and the digitized results are received and processed by the computer.
The computer examines the digital output of each analytical detector
channel at the beginning of every measurement period and, if necessary,
either increases or decreases the gain of the appropriate C–V converters
by up to two orders of magnitude. This auto-ranging process, which
requires only a few milliseconds at the beginning of each measurement
period, assures that the output of each of the C–V converters lies within
the voltage range (0.82–8.99 V) that yields maximum resolution for
digitization of the corresponding photocurrents. During the measurement period, each photomultiplier output channel is measured 1031
times/sec; the results of the individual measurements are averaged for
each channel at the conclusion of the measurement period. This data
acquisition approach is sometimes referred to as a digital time-sliced
average or digital time-sliced integration approach.

Wavelength Profiles. A second important feature of the polychromator system is that it has provisions for obtaining profiles of emission intensity vs. wavelength on a simultaneous multielement basis for many small wavelength regions, each of which encompasses one of the analytical lines of interest. This feature has been very valuable for diagnosis of instrumental stray light and spectral interferences (*16, 17*). In the present work, it provided the basis for our approach for ensuring accurate background correction. A complete description of the background correction approach is in preparation; thus, only a brief sketch will be given in the following.

When the angle of incidence of the plasma radiation falling on the grating is changed by moving the entrance slit of the polychromator a short distance along the focal curve, the angles of diffraction also change, and the entire dispersed spectrum moves along the focal curve in the opposite direction. If the entrance slit is moved from its usual analytical position to a position that is nearer (or farther from) the grating normal, the wavelengths sampled by the fixed exit slits no longer coincide with the wavelengths of the selected analytical lines but, instead, coincide with wavelengths that are all slightly less (or greater) than those of the respective analytical lines. If the entrance slit is moved at a constant rate through a range of positions that includes the usual analytical position, the strip-chart recording of the emission intensity detected at any single exit slit is comparable with the intensity vs. wavelength recordings that are commonly produced by rotation of the grating in a single-channel spectrometer. The recording for each exit slit includes the region of analyte emission and the adjoining regions of spectral background emission.

With the polychromator and a multichannel recorder, the intensity vs. wavelength profiles can be obtained for all of the analytical line wavelength regions simultaneously. Such a multichannel analog approach can be used for quantitative multielement analyses, but there is a more attractive alternative. Time-averaged or integrated intensity measurements can be obtained (simultaneously for each of the analytical lines) for each of a series of small, incremental changes in the position of the entrance slit. This approach yields low-noise digital profiles for all of the analytical line wavelength regions, simultaneously, and it is a practical approach for multielement analyses. Furthermore, if the incremental movements of the entrance slit are equivalent to wavelength changes of 0.005 nm or smaller, the shapes of the wavelength profiles can be approximated quite well by interpolation.

In our polychromator, the position of the entrance slit is determined by a micrometer screw that is driven by a stepper motor. The motor is operated by the minicomputer, hence the profiling operation is entirely automatic. The smallest wavelength change that can be accomplished with the stepping motor, 0.00026 nm, is approximately one tenth of the theoretical resolution of the polychromator. Therefore, most wavelength profiling work has been performed with wavelength increments that correspond to 30 or 40 motor steps, i.e., 0.0078 or 0.0104 nm. The total entrance slit movement for typical wavelength profiles performed in this work was approximately 0.3 mm ($-$ 0.12 nm). Therefore, with a 1:1 image of the plasma focussed at the entrance slit, the portions of the

Figure 4. A composite of intensity vs. wavelength profiles for copper in human urine and spiked urine samples. The reference blank solution was 1% (v/v) concentrated HCl in deionized distilled water (DDW).

(⊙) Reference blank, (+) urine, (×) urine + 10 ng/mL Cu, (◇) urine + 25 ng/mL Cu, (△) urine + 100 ng/mL Cu. The wavelength span of the profiles is approximately 0.12 nm; the entrance slit dial position (wavelength) that corresponds to the maximum intensity of the Cu 324.7 nm emission line (i.e., the usual analytical position) is indicated by a vertical arrow. The intensity (photocurrent) for each data point corresponds to the average for a 10-sec integration period.

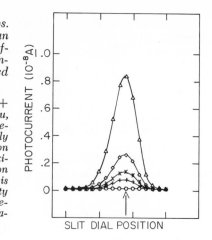

plasma that were observed at the extremes of each profile were separated by the same distance. The differences that could arise from the observation of different portions of the plasma within this small region at the center of the analytical zone of the plasma were not considered to be important because the maximum separation (0.3 mm) was much smaller than the width of the analytical zone. The latter width (distance between half-maximum intensity points) is approximately 5 mm (*18*). An example of the type of results that are obtained by wavelength profiling is shown in Figure 4.

Background Correction. The profiles shown in Figure 4 are simple because, in this case, the sample and reference blank solutions have the same background intensity, i.e., the profiles have identical emission intensities for the nonanalyte wavelength regions of the profiles. As was noted previously (*16, 17, 18, 19, 20*), this is not always so. Normal variations in the concentrations of concomitant elements in samples can cause the observed background intensities to be significantly different from those observed for the reference blank and can cause the background levels to vary from sample to sample by amounts which are often comparable with the quantities of trace analytes that are to be determined. As will be illustrated below, if these sample-to-sample background variations are unrecognized or ignored, serious analytical errors may arise.

The magnitude of the background problem is illustrated in Figure 5A for the case of selenium in urine. In Figure 5A, the differences in background intensity observed for the NaCl reference solutions and the urine "standard-addition" solutions (for example, at slit dial position −8.98, as indicated by the dashed arrow in the figure) are comparable with the net intensity which is attributable to the selenium content of the urine sample. If the conventional blank subtraction procedure for background correction were used for the determination of selenium in urine, i.e., using the difference between the intensity observed for the sample and that observed for the reference blank at the entrance slit dial position that would normally be used for the analysis (slit position

−8.82, as indicated by the solid arrow in the figure), the analytical result would be too high by approximately a factor of two. For the selenium example, it is apparent that the background correction problem is serious and that important analytical errors will result unless accurate correction procedures are used. Similar statements often apply for other elements as well, when the determinations must be performed for concentration levels near the limit of detection.

There are two main reasons for the variation of background intensities for different urine samples. First, the relatively high and variable concentrations of calcium and magnesium in urine can give rise to correspondingly high and variable amounts of stray light within the polychromator. Second, the magnesium present in urine also causes significant amounts of continuum-like emission which is a result of ion–electron recombination within the plasma (*19*). The stray light can be present at any of the exit slits, regardless of the wavelengths sampled. When it reaches the photomultipliers, the stray light increases the spectral background intensity underlying the corresponding analyte emission lines and can lead to analytical error. It is known that stray light

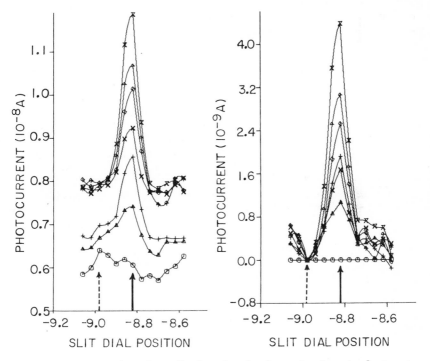

Figure 5. Wavelength profile data for the determination of selenium in urine: (A) (left) as observed, (B) (right) background-corrected profiles.

(⊙) *Reference blank: 0.5 wt % NaCl and 1% (vol) HCl, (△) reference blank + 50 ng Se/mL, (+) reference blank + 100 ng Se/mL, (×) normal urine sample (see Table IV), (◇) urine sample + 20 ng Se/mL, (⬟) urine sample + 50 ng Se/mL, (✕) urine sample + 100 ng Se/mL. The analytical line wavelength was 196.03 nm.*

contributions to the background can be reduced through the use of high quality, holographically recorded gratings; appropriate light-baffling and trapping within the instrument; and optical interference filters (17, 20, 21). The recombination emission, on the other hand, does not depend upon the quality of the instrumentation. It is a true spectral phenomenon, which may cause the background to vary from sample to sample, depending upon the concentration of concomitant elements, e.g., aluminum and magnesium.

For certain sample types, both stray light and recombination effects on the analytical results can be largely eliminated by matrix matching. Such an approach is not practical for the determination of trace elements in urine because of the wide ranges of calcium and magnesium concentrations which occur in so-called "normal" samples.

An alternative procedure for reducing the effects of stray light and recombination spectra upon the analytical results is to treat the effects as if they were caused by spectral line interferences (22). For this approach to be successful, the corrections that are applied must include terms for all those elements which contribute to significant background shifts. Apart from the problems associated with identifications of all of the "interfering" elements and with the determination of appropriate empirical equations for calculation of the correction terms, this approach has the additional complication that the individual background shift contributions may not be linearly related to either the intensity of any single emission line or to the concentration of an element that is responsible for a background shift (23). Another weakness of the interference–correction approach is that the accuracy and precision of the "corrected" analyte concentrations may be seriously impaired when a number of elements make significant contributions to the background at any given exit slit. Each of the concentration (or intensity) determinations required to ascertain the magnitude of the total "interference correction" for background shift makes its own independent contribution to the uncertainty of the final analytical result.

The approach presently used in our laboratory for avoiding the effects of sample-related background shifts (24) is based on the previously described wavelength-profiling capability. The spectral background at one or more preselected wavelengths in the immediate vicinity of each analyte line is measured directly for each sample. When such background information is used for the determination of net analyte intensities, the effects of sample-related stray light and recombination radiation are automatically eliminated. Typical results of the background correction procedure are illustrated in Figure 5B for the case of selenium in urine. The profiles for the individual solutions in Figure 5B are expressed in terms of net intensity (photocurrent). The entire profile for the reference blank solution has been subtracted from each of the other profiles, and the background shift with respect to the blank (determined from the observed intensity difference at slit dial position −8.98) also has been subtracted from each profile. After background correction, the profiles for selenium (Figure 5B) resemble the profiles which were shown earlier (Figure 4) as an example of experimental results that were not significantly affected by stray light or recombination emission.

Results and Discussion

The results shown in Figures 6 and 7 are typical of those obtained when the net intensities of the analyte lines were used as the measure of analytical response. (This is the usual procedure for ICP–AES analyses.) The curves for the urine samples with different total solids content exhibited different slopes, and none of them coincided with the curve observed for the (1% NaCl) reference solutions. The net intensities observed for the most concentrated solutions were inversely proportional to total solids content for both the urine samples and the reference solutions.

Figures 8 and 9, on the other hand, show typical analytical calibration curves for the same sample materials when the net intensity ratio (net analyte line intensity/net internal reference line intensity) was used as the measure of response. In this case, the curves exhibited a considerably smaller range of slopes for the various urine samples and were in much better agreement with the data for the reference solutions.

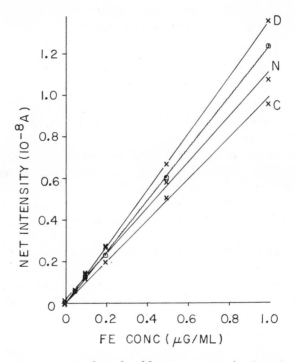

Figure 6. Analytical calibration curves for iron in 1% NaCl reference solutions (O) and in the dilute (D), normal (N), and concentrated (C) urine samples. See Table IV. The analysis wavelength was 261.2 nm.

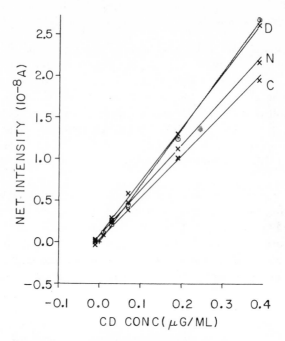

Figure 7. Analytical calibration curves for cadmium as in Figure 6. The analysis line wavelength was 226.5 nm.

Similar improvement in the correlation of the analytical curves was observed for all of the analytes when either the net intensity of the Ga(I) 294.3-nm line or that of the Y(II) 371.0-nm line was used for computing the intensity ratios. Figure 10, for example, shows the curves observed for Cr/Ga.

The improved concurrence of the analytical curves plotted in Figures 8, 9, and 10 indicates that the ratios of analyte to internal reference line intensities decreased the effects of differences in injection rate that were caused by changes in the amount of total dissolved solids. The fact that the net intensity of either the gallium or yttrium line served equally well for this "normalization" of the sample introduction rates indicates that, as expected, the differences in introduction rate had little effect on the actual excitation conditions within the plasma.

The data plotted in Figure 11 exemplify the typical results in a more revealing way. The slopes of the curves for analyte additions to the dilute, normal, and concentrated urine samples are nearly identical to each other and to the slope of the curve for the 1% NaCl reference solutions. Hence, urine samples with a wide range of total dissolved solids content can be analyzed successfully with only one set of (simple) reference solutions. Furthermore, the results of those determinations will be nearly identical to the corresponding analytical results from the method of additions. Also, the concentration intercepts of the dilute, normal, and concentrated urine addition curves yield nearly identical values for the zinc concentration of the composite (after consideration of the known dilution factors). As noted previously, this is exactly as it should be, because the urine samples all originated from the composite, and their zinc concentrations are therefore related to each other by those dilution factors.

Table V summarizes the quantitative results obtained for 13 trace elements in urine. Both the results from the method of additions and those obtained from the 1% NaCl analytical calibration curves are given. Concentrations were determined for dilute, normal, and concentrated urine solutions but, for ease of comparison, the results listed in the table are all reported in terms of the concentration of analyte present in the original composite urine sample. Background correction was performed for each sample and reference solution according to the wavelength-profiling procedure outlined above. A 10-second photocurrent measure-

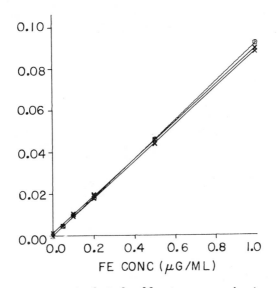

Figure 8.   Analytical calibration curves for iron as in Figure 6. Here, however, the analyte responses are in terms of the ratio of the net intensity of the analytical line (Fe 261.1 nm) to the net intensity of an internal reference element line (Y 371.0 nm). Although they are shown in the figure, the data points for the dilute, normal, and concentrated urine samples are not readily distinguishable.

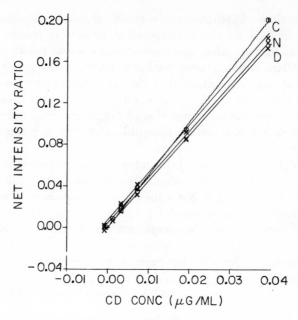

Figure 9.   Analytical calibration curves for cadmium as in Figure 8

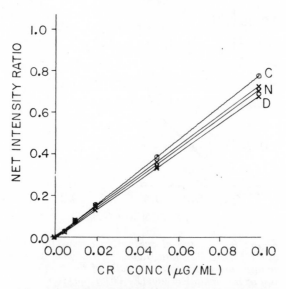

Figure 10. Analytical calibration curves for chromium as in Figures 8 and 9, but with gallium as the internal reference element. The analytical line was Cr(II) 205.6 nm; the internal reference line was Ga 294.4 nm.

ment was performed for each of 13 wavelengths (slit-dial positions) that defined the profile range, hence the analysis time, exclusive of calibration, was approximately five minutes per sample. The experimental limits of detection for the elements in the 1% NaCl matrix and in urine are reported in the last column of Table V.

Table V. Analytical Results for Trace Elements in Urine (ng/mL) [a]

Element	Dilute Urine	Normal Urine	Concentrated Urine	Det. Limit[b] (ng/mL)
Al	< 9	9	13	9
	< 12	< 12	9	12
As	< 70	100	< 70	70
	85	35	33	20
Be	0.07	0.06	0.07	0.03
	0.08	0.06	0.07	0.03
Cd	< 1	< 1	< 1	1
	< 5	< 5	< 5	5
Co	< 20	< 20	< 20	20
	< 10	< 10	< 10	10
Cr	< 1	< 1	< 1	1
	< 1	< 1	< 1	1
Fe	11	11	13	5
	11	10	12	3
Ni	< 9	< 9	< 9	9
	< 4	< 4	< 4	4
Pb	< 12	< 12	< 12	12
	< 15	< 15	< 15	15
Se	150	140	140	60
	120	130	110	80
Ti	< 10	< 10	< 10	10
	2	3	6	0.6
V	< 1	< 1	< 1	1
	< 1	< 1	< 1	1
Zn	500	530	500	50
	460	480	480	50

[a] Analytical results from the method of additions are given in the first row of results for each element. Those in the second row were obtained from the analytical calibration curve for the element in the 1% NaCl reference solutions.

[b] Concentration of analyte required to yield an average net intensity equal to three times the standard deviation of the background (blank) intensity (9). The upper entry for each element is the limit of detection for the analyte in 1% NaCl solutions; the lower entry for addition of analyte to the normal urine solutions.

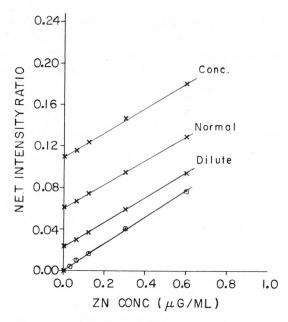

Figure 11. Analytical calibration curves for zinc in urine with yttrium as the internal reference element. The analytical and internal reference line wavelengths were 213.9 and 371.0 nm, respectively. (×) Additions to the urine samples, (◎) additions to the 1% NaCl reference soltuions.

Apart from the fact that some of the analyte elements were not detected at their normal concentration levels in urine, the main feature of the analytical results was their overall consistency. The results from the method of additions agreed well with those from the calibration approach and, in either case, the results obtained for the dilute, normal, and concentrated sample solutions also were in good agreement with each other. With the exception of arsenic and titanium, the results for the two methods and for the three urine concentrations were within one detection-limit concentration of each other for all of the elements. Although it has been noted (25) that detection-limit values are "inherently imprecise numbers" and that detection-limit concentrations "can only be detected, . . . , and not measured quantitatively," the consistency of the analytical results indicates that the background correction scheme was effective for elimination of the effects of stray light and recombination radiation. As noted earlier, the ratios of net analyte line to net internal reference line intensities were used to decrease the effects of sample-to-sample variations in total dissolved solids content.

The analytical results provided few surprises with respect to the trace element concentrations which were "expected" for urine. *See* Tables II and V. The results for all of the elements studied were within or below the range of previously reported concentrations. The selenium and zinc results were slightly greater than the corresponding "model" concentrations and those for aluminum, beryllium, cadmium, cobalt, chromium, iron, and vanadium were approximately one order of magnitude lower than the model values. Because of possible differences in diet and because the samples studied in this work were derived from first-morning voids, it is not possible to draw unequivocal conclusions concerning the differences between the observed and "expected" results.

The main disappointment that appears in Table V is that the experimental limits of detection for the analytes in the 1% NaCl matrix and in the urine samples were approximately one order of magnitude greater than had been observed previously for 1% HNO_3 solutions and for solutions containing both 1% NaCl and 1% HNO_3 (2). We are at a loss to explain the apparent conflict of these observations. The unexpectedly high limits of detection substantially reduced our expected capability for the quantitative determination of the normal levels of these analytes in urine.

Although the determination of boron, copper, and manganese and of the macroelements, calcium, magnesium, and phosphorus, was also studied, no results are reported for these elements. The macroelement investigation was rather cursory because the determination of these elements was not of primary interest. Nevertheless, the preliminary results showed that ICP–AES is applicable for the determination of calcium, magnesium, and phosphorus in urine. The determination of boron, copper, and manganese was studied in greater detail, but analyte volatilization and memory effects associated with aerosol desolvation made the intensity data for boron and copper unreliable; a (suddenly) noisy photomultiplier on the manganese channel had a similar effect on the intensity results for that element. The manganese problem was purely instrumental and has an easy solution. (Our preliminary experiments yielded a detection limit of approximately 3 ng/mL for manganese in urine and indicated that the "normal" manganese level was less than that value.) For the determination of boron and copper, a longer element cleanout period (i.e., > 2 min) and a lower desolvation temperature may be required. Also, a sample preparation procedure which would prevent the formation of volatile chlorides may be more suitable for these elements than the procedure used in the present work.

The principal conclusions of the present study are as follows. First, the ICP–AES technique is useful for the rapid, simultaneous determination of trace elements in urine. However, the limits of detection will have to be improved or sample preparation procedures that provide some preconcentration of the analytes prior to their introduction into the plasma will have to be developed before some of the elements of interest can be determined at their normal concentration levels. Second, the analytical errors which can arise from stray light and from ion–electron recombination emission can be eliminated if accurate background correction procedures are used. Third, the internal reference principle is useful for the "normalization" of sample introduction rates. The necessary control operations and calculations for proper background correction and for use of the internal reference approach are well within the capabilities of the computers which are often included in present, commercially available, ICP–AES instrument systems.

Acknowledgment

This investigation was supported by the National Institute for Occupational Safety and Health under Interagency Agreement NIOSH-IA-76-23 and was based on initial research supported by the Department of Energy, Office of Basic Energy Sciences, Chemical Sciences Subactivity.

Literature Cited

1. Fassel, V. A., Kniseley, R. N., *Anal. Chem.* (1974) **46**, 1110A.
2. Olson, K. W., Haas, W. J., Jr., Fassel, V. A., *Anal. Chem.* (1977) **49**, 632.
3. Larson, G. F., Fassel, V. A., Scott, R. H., Kniseley, R. N., *Anal. Chem.* (1975) **47**, 238.
4. Larson, G. F., Fassel, V. A., *Anal. Chem.* (1976) **48**, 1161.
5. Boumans, P. W. J. M., deBoer, F. J., Dahmen, F. J., Hoelzel, H., Meier, A., *Spectrochim. Acta* (1975) **30B**, 449.
6. *Report of the Task Group on Reference Man*, International Commission on Radiological Protection, Report # **23**, Pergamon, New York, 1975.
7. Mertz, W., "Trace Element Analysis in Nutrition Research," *in* "Abstracts of Papers," 174th National Meeting, ACS, Chicago, August 1977, ANAL 008.
8. Parr, R. M., "Intercomparison of Trace and Other Elements in IAEA Animal Muscle, H-4," Working paper # AG-51/6, IAEA Advisory Group on Optimal Methods for the Assay of Trace Elements in Biological Materials, Vienna, Nov.–Dec., 1976.
9. Commission on Spectrochemical and Other Procedures for Analysis, Analytical Chemistry Division, International Union of Pure and Applied Chemistry, *Pure Appl. Chem.* (1976) **45**, 99. (Also reprinted in *Anal. Chem.* (1976) **48**, 2294.
10. Fassel, V. A., Kniseley, R. N., *Anal. Chem.* (1974) **46**, 1155A.
11. Winge, R. K., Fassel, V. A., Kniseley, R. N., DeKalb, E., Haas, W. J., Jr., *Spectrochim. Acta* (1977) **32B**, 327.
12. Menis, O., Rains, T. C., "Sensitivity, Detection Limit, Precision and Accuracy in Flame Emission and Atomic Absorption Spectrometry," *in* "Analytical Flame Spectroscopy," R. Mavrodineanu, Ed., pp. 73–74, MacMillan, London, 1970.
13. Robertson, D. E., *Anal. Chim. Acta.* (1968) **42**, 553.

14. Ahrens, L. H., Taylor, S. R., "Spectrochemical Analysis," 2nd ed., Addison–Wesley, Reading, MA, 1961.
15. Barnett, W. B., Fassel, V. A., Kniseley, R. N., *Spectrochim. Acta* (1968) **32B**, 643.
16. Haas, W. J., Jr., Winge, R. K., Fassel, V. A., Kniseley, R. N., "Trace Level Multielement Analysis by Atomic Emission Spectrometry. Applications of a Simultaneous Multielement Wavelength Profiling Facility for Diagnosis of Stray Light and Spectral Line Interference Effects," Third Annual Meeting of the Federation of Analytical Chemistry and Spectroscopy Societies, Philadelphia, PA, November 15, 1976, paper no. **79**.
17. Larson, G. F., Fassel, V. A., Winge, R. K., Kniseley, R. N., *Appl. Spectrosc.* (1976) **30**, 384.
18. Winge, R. K., Fassel, V. A., Development and Application of an Inductively Coupled Plasma Analytical System for the Simultaneous Multielement Determination of Trace Elemental Pollutants in Water," Annual progress report submitted to the Southeast Environmental Research Laboratory, U.S. EPA, Athens, GA, 1975.
19. Larson, G. F., Fassel, V. A., "Line Broadening and Radiative Recombination Background Interference for Inductively Coupled Plasma Excitation," manuscript in preparation.
20. Abercrombie, F. N., Silvester, M. D., Cruz, R. B., "Simultaneous Multielement Analysis of Biologically Related Samples with a RF-ICP Advantages and Limitations," *in* "Abstracts of Papers," 174th National Meeting, ACS, Chicago, August 1977, ANAL 009.
21. Fassel, V. A., Katzenberger, J. M., Winge, R. K., "Effectiveness of Interference Filters for Reducing Stray Light Effects in Atomic Emission Spectrometry," accepted for publication, *Appl. Spectrosc.* (1979).
22. Dalhquist, R. L., Knoll, J. W., *Appl. Spectrosc.* (1978) **32**, 1.
23. Winge, R. K., Katzenberger, J. M., Fassel, V. A., "Development and Application of an Inductively Coupled Plasma Analytical System for the Simultaneous Multielement Determination of Trace Elemental Pollutants in Water," Annual progress report submitted to the Southeast Environmental Research Laboratory, U.S. EPA, Athens, GA, 1977.
24. Haas, W. J., Jr., Fassel, V. A., unpublished data.
25. O'Haver, T. C., "Analytical Considerations," *in* "Trace Analysis, Spectroscopic Methods for Elements," J. D. Winefordner, Ed., Wiley, New York, 1976.

RECEIVED June 30, 1978.

9

The Role of Zinc in Biochemical Processes

D. S. AULD

Biophysics Research Laboratory, Department of Biological Chemistry,
Harvard Medical School and the Division of Medical Biology,
Peter Bent Brigham Hospital, Boston, MA 02115

Analyses of in situ DNA synthesis of Euglena gracilis *identify zinc-dependent steps in the eukaryotic cell cycle and show that the derangements in RNA metabolism are critical determinants of the growth arrest associated with zinc deficiency. Combined use of microwave-induced emission spectrometry and micro gel emulsion chromatography shows the presence of stoichiometric amounts of zinc essential to the function of E. gracilis and yeast RNA polymerases, the "reverse transcriptases" from avian myeloblastosis, murine leukemic and woolly type C viruses, and E. coli methionyl tRNA synthetase. These results stress the importance of zinc to both nucleic acid and protein metabolism. Transient-state kinetic studies of carboxypeptidase A show that zinc functions in the catalytic step of peptide hydrolysis and in the binding step of ester hydrolysis.*

Evidence for Biochemical Role of Zinc

The first report that zinc is essential to growth was made more than one hundred years ago. In 1869, Raulin found that zinc was indispensable to the growth of *Aspergillus niger* (*1*), and its presence in plants and animals (*2, 3*) was demonstrated within a decade (Table I). During the next half century, biological science advanced rapidly but the biological role of zinc remained virtually unknown. It was not until 1934 that conclusive evidence was given that zinc is essential to the normal growth of rodents (*4*), a fact now appreciated to pertain to all forms of life. Since then an increasing number of diseases have proved to be caused by zinc deficiency or imbalance. Quite recently, acrodermatitis enterohepatica, a fatal, congenital disease, has been cured by administration of this metal (*5*).

0-8412-0416-0/79/33-172-112$05.50/0

Table I.   **Evidence for Biological Roles of Zinc (1869–1955)**

Raulin	Zn deficiency in *A. niger*	1869
Lechartier and Bellamy	Zn in plants and vertebrates	1877
Raoult and Breton	Zn in liver	1877
Bertrand	Zn deficiency in rodents	1934
Keilin and Mann	Zn carbonic anhydrase	1940
Vallee and Neurath	Zn carboxypeptidase	1955

One of the most important discoveries concerning the biological role of zinc occurred in 1940 when Keilin and Mann showed that zinc is an essential compound of erythrocyte carbonic anhydrase, an enzyme catalytically involved in the transport of CO_2 in blood (6). Following the 70-year interval between the initial recognition of a metabolic zinc deficiency and the characterization of the first zinc metalloenzyme, there was a period of about 15 years before the second zinc enzyme was identified. In 1955, Vallee and Neurath reported that carboxypeptidase A from bovine pancreas contained 1 g-atom Zn per mol of protein and was essential to the function of the enzyme (7). The presence of zinc in carbonic anhydrase and carboxypeptidase A indicated that a primary role of zinc would be to function in zinc metalloenzymes (62). However, it seemed unlikely that disrupting the activity of carboxypeptidase A or carbonic anhydrase would have profound effects on growth.

The ubiquity of zinc metalloenzymes has been appreciated only in the last decade, and rather suddenly their systematic study has become a significant area of biochemical inquiry. It would seem that in large measure the lack of visible color of zinc proteins can be held accountable for the long delay in their recognition while, in contrast, the red iron (8) and blue copper (9) proteins early called attention to themselves (10). In fact, the isolation, purification, and recognition in 1940 of carbonic anhydrase as a zinc enzyme (6) was a happy accident. Through much of their purification, Keilin and Mann's attention was directed to a blue protein, apparently exhibiting carbonic anhydrase activity, and hence it was suspected to be a copper enzyme. This blue protein actually proved to be hematocuprein, now recognized as superoxide dismutase, an enzyme that contains both copper and zinc. However, as a result of only minor adjustments of conditions, the analytical method used for copper could also detect zinc, leading to the recognition of carbonic anhydrase as the first zinc metalloenzyme.

Zinc and Eukaryotic Cellular Metabolism

To study zinc at the cellular level, in vivo, an organism must be available that can be grown rapidly under rigorously controlled trace metal conditions, can be obtained in homogeneous form, and can be readily disrupted to allow unambiguous subcellular analysis. The alga, *Euglena gracilis*, was found to satisfy these critical prerequisites and hence has been used as a model to study the biochemical and morphological consequences of zinc deprivations (*11, 12, 13, 14, 15*).

Under normal conditions of zinc sufficiency ($10^{-5}M$), *E. gracilis* cultures reveal a 5–6 day lag period followed by a period of logarithmic growth which continues through days 12–14 (Figure 1). The cells then enter a stationary phase, and there is no further increase in total cell number. If zinc is limiting ($10^{-7}M$), there is a marked reduction in cellular proliferation which is most noticeable after one week of incubation (Figure 1). If zinc, and only zinc, is added to the deficient media, cell growth ensues within 24–48 hr and after a few days reaches the same level as in the initially zinc-sufficient media. Hence, growth impairment is caused solely by a deficiency of zinc. An analysis of the zinc-sufficient and -deficient cells after 13 days of culturing reveals sevenfold greater zinc content (58 vs. 8 μg Zn/10^8 cells) in the former (*12, 15*).

A number of striking morphological and chemical changes accompany zinc deficiency-induced growth arrest. Cell volume and size increase while osmophilic granules and paramylon accumulate; cellular DNA

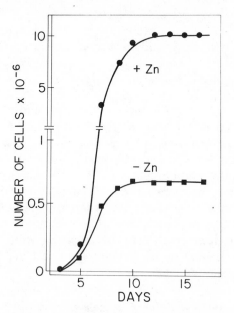

Figure 1. Growth of zinc-sufficient (●) and zinc-deficient (■) E. gracilis grown in the dark. Zinc-sufficient medium contains 1×10^{-5}M Zn^{2+}; zinc deficient medium contains 1×10^{-7}M Zn^{2+} (12).

content doubles; RNA content remains constant, protein synthesis is depressed; and peptides, amino acids, nucleotides, polyphosphates, and unusual proteins accumulate, suggesting that zinc is importantly involved in nucleic acid metabolism and cellular division. Indeed, the rate of incorporation of [³H] uridine into RNA of zinc-deficient *E. gracilis* is decreased (*12*), the element maintains the integrity of ribosomes of these organisms, and it seems to be required for the stability of both RNA and DNA.

To better define the specific lesions that accompany zinc deficiency, the events occurring during the typical eukaryotic cell cycle have been examined. *E. gracilis*, though a eukaryote, undergoes the growth phases usually associated with prokaryotic organisms, i.e., lag, log, and stationary phases. However, these are analogous to the more detailed G_1, S, G_2, and mitosis phases (*16*). *E. gracilis* cells increase in size and synthesize RNA and protein during G_1 (*13*). In S, increased DNA synthesis results in chromosomal replication. The mitotic apparatus forms during G_2, followed by nuclear and cellular division.

Recent laser excitation flow cytofluorometry of the dynamics of DNA metabolism in the cell cycle in zinc-deficient *E. gracilis* leave no doubt of the involvement of zinc in cell replication (*13*). The data demonstrate that all of the biochemical processes essential for cells to pass from G_1 into S to G_2 and from G_2 to mitosis require zinc, and its deficiency can block all phases of the growth cycle of this organism.

Zinc and RNA-Dependent DNA Polymerases

Recognition of the existence of an RNA-dependent DNA polymerase —a reverse transcriptase—in avian myeloblastosis viruses (*17, 18*) has given new insight into the biochemical basis of leukemic and other malignant transformations. The importance of zinc in the metabolism of normal and leukemic leukocytes (*19, 20*) prompted us to examine the RNA-dependent DNA polymerase from avian myeloblastosis virus for its metal content (*21, 22*). Minimally, this required the analytical demonstration that a metal atom is present. The paucity of available material presents formidable problems for such an analysis, but the marked inhibition of enzyme activity by metal-chelating agents (*21, 22*) led us to devise a system capable of quantitative metal determinations at the 10^{-13} g level. Microwave-induced emission spectrometry provided the microanalytical system capable of precisely measuring 10^{-10}–10^{-13} g of metal in microgram amounts of enzyme, orders of magnitude more sensitive than other, more conventional methods. Extensive studies of the effect of various cations, anions, and organic molecules on the emission lines of metals determined by this method have allowed the definition of optimal conditions for analysis (23). There is a tenfold to 1000-fold enhancement in sensitivity if analyses for metals are performed in the presence of potassium chloride (4–10 mmol/L). In the presence of KCl, the limit of zinc detection is lowered to 0.3 $\mu g/L$ ($\sim 2 \times 10^{-13}$ g for a 5-μL sample).

The accuracy of the microwave-excitation spectrometric method was verified by comparing results from it with those of atomic absorption analysis for readily available metalloenzymes of known zinc stoichiometry. Carboxypeptidase A (EC 3.4.12.2), carbonic anhydrase (EC 4.2.1.1), alcohol dehydrogenase (EC 1.1.1.1), and alkaline phosphatase (EC 3.1.3.1) were dialyzed vs. metal-free buffers, then diluted with 10 mmol/L KCl or 1 mmol/L HCl for metal analysis (24). For atomic absorption analysis, at least 100-μg samples were required, but microwave excitation required only 0.1 μg. Even though 1000-fold less protein was required for microwave excitation analysis, the agreement between the data obtained by the two methods is excellent (Table II). So little of the reverse transcriptase was available to us that we could not use atomic absorption for its analyses.

The RNA and DNA dependent polymerases are normally purified in buffered solutions containing 25–50% glycerol, 0.2 to 0.3M KCl, Triton X-100, and millimolar concentrations of DTT or mercaptoethanol. Because matrix effects are numerous under such conditions, it is important to separate proteins from the contaminating small molecules that might interfere with metal analysis. The removal of these contaminants is accomplished by combining microscale gel exclusion chromatography with microwave-induced emission spectrometry (24). A schematic of the experimental arrangement is shown in Figure 2. A 1.5 mL microbore

Table II. Comparison of Microwave-Induced Emission and Atomic Absorption Spectrometric Determination of Zinc Stoichiometry in Known Zinc Metalloenzymes[a]

Enzyme	Microwave Emission	Atomic Absorption
	(g-atom Zn/mol of protein)	
Bovine carboxypeptidase A	1.0	1.0
Human carbonic anhydrase	1.1	1.1
Horse liver alcohol dehydrogenase	4.2	3.9
E. coli alkaline phosphatase	3.7	3.6
Reverse transcriptases		
AMV[b]	1.9	—
Woolly Monkey	1.4	—
Murine	1.0	—

[a] Zinc content determined in ~ 0.1 μg of protein for microwave-induced emission spectrometry and ~ 0.1 mg for atomic absorption spectrometry.
[b] Avian myeloblastosis virus.

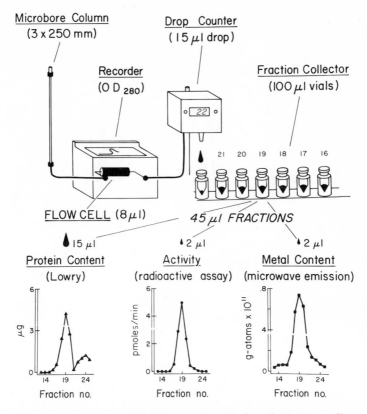

Figure 2.   Microanalytical system for metal analysis of metallo-enzymes

column (0.4 × 25 cm) of Sephadex G-100 was purified of contaminating metal ions by washing with 1,10-phenanthroline and was used for gel exclusion chromatography. The effluent from the column then passes through an 8 μL flow-through cell and the optical density is recorded at 280 nm. Droplets of effluent are counted by a 15 μL drop counter. The 45 μL fractions are collected in Kontes 100 μL reaction vials which are seated in a Gilson escargot fraction collector. In this manner all functions after addition of the sample to the column are automatic, and a direct nondestructive monitoring of the protein is achieved. The fractions then can be analyzed for protein by a micro-Lowry technique enzyme activity by measuring the incorporation of [methyl-³H]thymidine-5'-monophosphate into an acid precipitable nucleic acid homopolymer and metal content by microwave-induced emission spectroscopy.

The results for the avian enzyme are shown in Figure 3. Fifty μL containing 50 μg of partially purified avian enzyme were applied to the column. The 45-μL droplet fractions were measured for protein content,

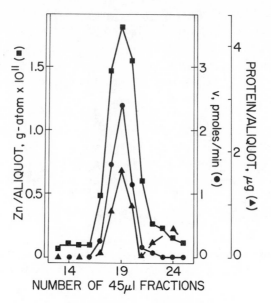

Figure 3. Distribution of AMV polymerase activity (●), zinc (■), and protein (▲) from a G-100 Sephadex column. After metal ions were removed with 1,10-phenanthroline, the column was equilibrated at 4°C with a mixture of KCl (10 mmol/L), tris(hydroxymethyl)aminometh-ane (pH = 7.8, 10 mmol/L), dithiothreitol (1 mmol/L), and Triton X-100 (10 μL/L). A 50 μL sample of the AMV enzymes was placed on the column and eluted with the above buffer. Droplet fractions, 45 μL, were collected for assays in duplicate and zinc analyses in tripli-cate. The zinc content is represented as g-atom per 5-μL aliquot (1 g-atom is equivalent to 64.5 g of zinc). The velocity is expressed for 1 μL of enzyme added to 100 μL assay con-taining: $MnCl_2$ (0.2 mmol/L), polyadenylic acid (0.1 μmol/L), oligodeoxythymidylic acid (1 μmol/L), [methyl-^{3}H]thymidine-5-triphos-phate (2.4 μmol/L), dithiothreitol (2 mmol/L), KCl (80 mmol/L), tris(hydroxymethyl)amino-methane (pH = 7.8, 0.11 mol/L).

enzyme activity, and zinc. One major peak of protein, enzyme activity, and zinc is obtained centered about fraction 19. In the most active frac-tions, 18, 19, and 20, the zinc:activity and zinc:protein ratios are virtually constant, indicating purification concomitant with gel exclusion chroma-tography. The elution patterns are quite reproducible. Fractions 18, 19, and 20 were found to contain 1.5, 1.8, and 1.6×10^{-11} g-atom of zinc per aliquot, corresponding to a ratio of 1.8 to 1.9 g-atom of zinc per mol of protein of 150,000 molecular weight. Analyses for copper, iron, and manganese were also carried out for each fraction. These analyses demonstrated that, summed together, these elements were present in insignificant amounts. Thus, AMV reverse transcriptase is a zinc metallo-enzyme containing 2 g-atom of zinc, which may correspond to the two known subunits. These results led us to inspect reverse transcriptases from the woolly monkey and murine sources in a similar fashion (25), and the results demonstrate these enzymes contain 1.4 and 1.0 g-atom of zinc respectively (Table II). The lower zinc content of the mammalian enzymes might reflect the fact that they are monomeric enzymes.

Zinc in Nucleic Acid and Protein Metabolism

We also have used the microanalytical system (Figure 2) for examining the metal content of a number of eukaryotic RNA polymerases (26, 27, 28). The results for RNA polymerase I from yeast are shown in Figure 4 (26). As shown previously for the reverse transcriptases, the droplet fractions were measured for protein and zinc content and for enzyme activity. The results show one major peak of protein/enzyme

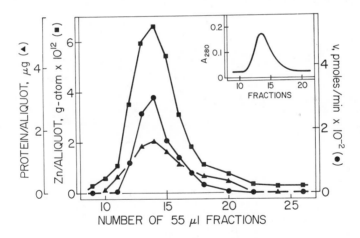

Biochemical and Biophysical Research Communications

Figure 4.   Distribution of yeast RNA polymerase I activity (●), zinc (■), and protein (▲) in 55 μL fractions from a G-75 Sephadex column (26). Metal contaminants were removed from the column and the column equilibrated with buffer as described in Figure 3. Approximately 50 μg of protein in a 50 μL sample was placed on the column and fractions were collected by elution with buffer at a flow rate of 50 μL/min. The insert is the optical density recorded at 280 nm of the effluent of the column measured in 8 μL flow through cell.

activity and zinc is obtained centered about fraction 14. The inset on the right shows the direct absorbance reading at 280 nm. The absorbance closely follows the other three determinations. The direct absorbance reading should make it much simpler to follow protein in future studies. Fractions 13, 14, and 15 were found to contain 1.5, 1.6, and 1.3 μg of protein and 5.9, 6.6, and 5.4 $\times$ 10^{-12} g-atom of zinc per aliquot, respectively, corresponding to a ratio of 2.4 g-atom of zinc per mol of protein. Analyses for copper, iron, manganese, and magnesium demonstrated that together these elements were present in insignificant amounts. It is interesting to note that this polymerase, like the reverse transcriptases, while needing either manganese or magnesium to function does not contain either of these metals in the holoenzyme. The function of these metal ions, Me^{2+}, can be to interact with the nucleotide substrate. Zinc, on the other hand, is tightly bound and inhibition by metal binding agents suggests it has a catalytic role.

Further support for the involvement of zinc in eukaryotic RNA synthesis comes from the study of *E. gracilis* grown under zinc-sufficient conditions. Both the RNA polymerases I and II have been purified to homogeneity (27, 28). These enzymes contain on the average 2.2 g-atom of zinc per mol of protein and are inhibited by the metal chelator 1,10-phenanthroline. The DNA-dependent DNA polymerase of *E. gracilis* has been isolated recently and is inhibited by 1,10-phenanthroline (29), suggesting it too can be a zinc-containing enzyme.

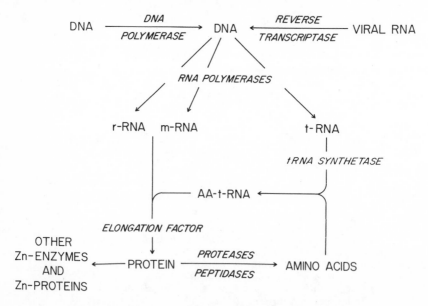

Figure 5.   Zinc in nucleic acid and protein metabolism

Zinc also has been demonstrated to be essential to the function of protein synthesis elongation factor 1 (Light Form EF_L) from rat liver which catalyzes the binding of aminoacyl tRNA to an RNA ribosome site through the formation of an aminoacyl tRNA · EF_L · GTP ternary complex (30). Most recently we have established that zinc is present in s-Met tRNA synthetase from *E. coli* (31). Figure 5 briefly summarizes the importance of zinc to nucleic acid and protein metabolism. The effect of zinc on cellular growth and development and in particular on the different stages of the eukaryotic cell cycle is not unreasonable in view of the unraveling picture of the importance of zinc to nucleic acid and protein metabolism. The present data provide evidence for a role of zinc in transcription and translation of eukaryotes, prokaryotes, and viruses underlying the essentiality of zinc in cell division. The demonstration that eukaryotic RNA polymerase I and II (26, 27, 28) are zinc enzymes support the hypothesis that the element is essential for RNA metabolism in all phyla (21, 24–28, 32–35). The data further suggest that extension of our studies on zinc-deficient *E. gracilis* to other organisms can assist in generalizing the emerging biochemical role of zinc in growth, proliferation, and differentiation (10).

Zinc Metalloenzymes (1977)

The increased rate at which zinc metalloenzymes have been identified in the past decade has been caused by, in large measure, the development of highly precise, rapid, and convenient methods of analyzing for zinc. While in 1967 there were only about one dozen zinc enzymes identified, there are presently more than 90 (Table III). Zinc is now known to participate in a variety of metabolic processes, including carbohydrate, lipid, protein, and nucleic acid synthesis or degradation. It is essential for the function and/or structure of at least one in each of the six categories of enzymes designated by the Commission on Enzyme Nomenclature of the International Union of Biochemistry and present throughout all phyla (Table III), among them several dehydrogenases, aldolases, peptidases, phosphatases, an isomerase, a transphosphorylase, and aspartate transcarbamylase. The functional characteristics of zinc metalloenzymes are remarkably diverse, and the role played by the metal atom is not yet established for each such enzyme.

Chemical Features of the Zinc Enzyme

Zinc does not undergo a change in oxidation state during enzymatic catalysis even though it participates in oxidoreduction reactions, e.g., as a component of alcohol dehydrogenase. The zinc cation has a stable,

Table III. Zinc Metalloenzymes, 1977

Enzyme	No.[a]	Source
Alcohol dehydrogenase	5	plants, vertebrates
D-Lactate cytochrome reductase	1	yeast
Superoxide dismutase	6	plants, vertebrates
Aspartate transcarbamylase	1	*E. coli*
Transcarboxylase	1	*P. shermanii*
Phosphoglucomutase	1	yeast
RNA polymerase I and II	2	*E. gracilis*
RNA polymerase I and II	2	yeast
RNA polymerase II	1	wheat germ
DNA polymerase I and II	2	*E. coli*, sea urchin
Reverse transcriptase	5	oncogenic viruses
Mercaptopyruvate sulfur transferase	1	*E. coli*
Alkaline phosphatase	5	*E. coli*, vertebrates (intestine, placenta)
Phospholipase C	1	*B. cereus*
Leucine aminopeptidase	4	kidney, lens, yeast
Particulate leucine aminopeptidase	1	kidney
Carboxypeptidase A	4	pancreas
Carboxypeptidase B	2	pancreas
Collagenase	1	*Clostridium histolyticum*
Procarboxypeptidase A	2	pancreas
Procarboxypeptidase B	1	pancreas
Carboxypeptidase G1	1	*P. stutzeri*
Dipeptidase	2	kidney, mouse tumor
Neutral protease	7	bacteria
AMP aminohydrolase	1	muscle
Aldolase	2	yeast, *A. niger*
L-Rhamnulose 1-phosphate aldolase	1	*E. coli*
Carbonic anhydrase	18	erythrocytes, muscles, plants
δ-Aminolevulinic acid dehydratase	1	liver
Phosphomannose isomerase	1	yeast
Pyruvate carboxylase	1	yeast
Protein elongation factor	1	liver
α-D-Mannosidase	1	jack bean
β-Lactamase I and II	2	*B. cereus*
t RNA synthetase	2	*E. coli*
Deoxynucleotidyl transferase	1	calf thymus

[a] Number of species from which enzyme was isolated.

d_{10} electronic configuration and has little tendency to accept or to donate single electrons. Instead, it likely serves as a Lewis acid interacting with electronegative donors to increase the polarity of chemical bonds and thus promote the transfer of atoms or groups (*36*). Substitution reactions of simple metal chelates generally proceed via intermediates with an open coordination position or a distorted coordination sphere. Zinc can readily accept a distorted geometry and, hence, would appear to be well suited to participate in substitution reactions as, for instance, in carbonic anhydrase, carboxypeptidase, and alkaline phosphatase.

The three-dimensional structure of proteins, heterogeneity of ligands, and the degree of vicinal polarity of the metal binding site can jointly generate atypical coordination properties (*37*). Unusual bond lengths, distorted geometries, and/or an odd number of ligands can generate a metal binding site on the enzyme, which when occupied by a metal, can be thermodynamically more energetic than metal ions when free in solution, where they are complexed to water or to simple ligands. As a result, zinc in enzymes is thought to be poised for its intended catalytic function in the entatic state (*37*). In this context, the term entasis indicates the existence of a condition of tension or stress in a zinc (or other metal) enzyme prior to combining with substrate.

Zinc enzymes offer unusual opportunities to study the manner in which specific function is achieved through interaction of the metal with the protein. In a number of cases, zinc can be replaced with other metals with retention of some degree of activity. Replacement of it with paramagnetic metals, e.g., cobalt, can signal information on absorption (*38, 39*), magnetic circular dichroic (*40*), and electron paramagnetic resonance spectra (*41*), combined with kinetic studies of cobalt-substituted and chemically modified zinc metalloenzymes, have enlarged understanding both of the possible modes of interaction of zinc with the active sites of zinc metalloenzymes and of their potential mechanisms of action.

The molecular details of the action of metalloenzymes have begun to be elucidated in the past few years (*42*). Crystal structures for bovine carboxypeptidase A (*43*), thermolysin (*44*), and horse liver alcohol dehydrogenase (*45*) are now available, and chemical and kinetic studies have defined the role of zinc in substrate binding and catalysis. In fact, many of the significant features elucidating the mode of action of enzymes in general have been defined at the hands of zinc metalloenzymes.

The Role of Zinc in Carboxypeptidase A

Carboxypeptidase A is one of the most intensely investigated zinc metalloenzymes. The enzyme as isolated contains 1 g-atom of zinc per protein molecular weight of 34,600. Removal of the metal atom either by dialysis at low pH or by treatment with chelating agents gives a totally inactive apoenzyme (*46*). Activity can be restored by readdition of zinc or one of a number of other di-valent metal ions (*47*). Through a combined use of chemical modification and transient state kinetic studies, it has been possible to determine the role of zinc in the catalysis of ester and peptide hydrolysis by this enzyme.

The past decade has witnessed a lively debate regarding the mechanisms of the carboxypeptidase-catalyzed hydrolysis of esters and peptides. The question of whether the mechanisms are the same or different has resisted definitive solution. A large body of information is available on peptide hydrolysis, but infinitely much less is known about that of their exact ester analogs. However, we have recently synthesized esters (48, 49, 50) which are exact structural analogues of amino-blocked oligopeptides (51). The hydrolysis of these new esters, like their oligopeptide analogs, conform to Michaelis–Menten kinetics (51, 52, 53). In both instances, fluorescent N-terminal blocking groups can be used, so that enzyme–substrate complexes can be observed (48, 54, 55, 56, 57). These substrates have the general form:

$$R—(Gly)_n—X—\overset{\overset{\textstyle R'}{|}}{C}H—COOH$$

where X is O or NH. For the peptide the C-terminal residue is an L-amino acid while for the ester it is the corresponding β-substituted L-lactic acid. The exact ester analog for the peptide substrate containing the phenylalanyl residue (X = NH, R' = CH_2Ph) is therefore the L-phenyllactyl derivative (X = O, R' = CH_2Ph). Thus, each ester–peptide pair differs only by virtue of the susceptible bond. The N-terminal blocking group, R, is either the fluorescent dansyl group or the conventional benzoyl or carbobenzoxy groups. The length of the substrate varies with the number, n, of glycyl residues (48, 57).

Resonance energy transfer between fluorescent enzyme trpytophanyl residues as intrinsic donors and the extrinsically placed acceptor, the dansyl group, in the substrate allows direct visualization of the enzyme–substrate complex (54, 57). The spectral overlap between the dansyl group absorption and tryptophan emission is excellent, and the dansyl emission spectrum is red shifted far enough not to overlap with its own absorption spectrum (Figure 6), properties that make these an exceptionally good donor–acceptor pair. Quantitatively, the degree of energy transfer is sensitive to the distance between and orientation of the donor–acceptor pair and to the environment of the acceptor. Differences in tryptophan-to-dansyl transfer efficiencies and/or dansyl yields can characterize the ES species formed (57).

The oscilloscope tracing in Figure 7 demonstrates the rapid enhancement of dansyl fluorescence following mixing of $1 \times 10^{-4}M$ Dns-$(Gly)_3$-L-OPhe and $2.5 \times 10^{-6}M$ carboxypeptidase A and reflects the extremely rapid equilibration of enzyme and substrate to form the ES complex. The decrease in the signal is a considerably slower process and, as

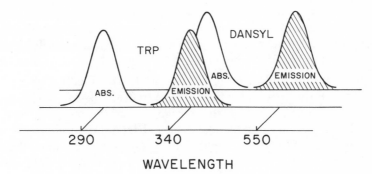

Enzyme Action

Figure 6.   Representation of the spectral overlap relationships between the enzyme tryptophan and substrate dansyl groups, which constitute the energy donor–acceptor pair critical to observation of the ES complex (57)

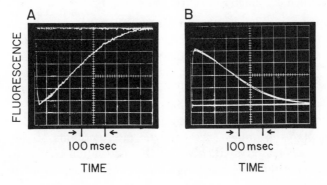

Enzyme Action

Figure 7.   (A) Enzyme tryptophan and (B) substrate dansyl fluorescence during the time course of hydrolysis of Dns-(Gly)₃-L-OPhe, 1 × 10⁻⁴M, catalyzed by zinc carboxypeptidase, 2.5 × 10⁻⁶M, in 1M NaCl–0.03M Tris, pH = 7.5, 25°C (57). The fluorescence of either (A) tryptophan or (B) dansyl was measured as a function of time under stopped-flow conditions. Oscilloscope traces of duplicate reactions are shown in each case. Excitation was at 285 nm. Enzyme tryptophan fluorescence was measured by means of band-pass filter peaking at 360 nm, and dansyl emission was measured by a 430-nm cutoff filter. Scale sensitivities for (A) and (B) are 50 and 500 mV/div, respectively. The existence of the ES complex is signaled by either (A) the suppression of enzyme tryptophan fluorescence (quenching by the dansyl group) or (B) enhancement of the substrate dansyl group fluorescence (energy transfer from enzyme tryptophan).

hydrolysis reduces the ester concentration, reflects a concomitant diminution in the concentration of enzyme-bound ester. The complementary pattern is observed when quenching of enzyme–tryptophan fluorescence by the dansyl group of the bound ester is measured.

The maximal fluorescence, F_{max}, reflects the binding strength of substrates and the area under the curve, A, reflects the catalytic rate (56, 57). Direct examination of the enzyme–substrate complex therefore allows determination of individual rate and equilibrium constants for a given substrate (57), direct deduction of modes of inhibition (56, 57), and the mechanistic consequences of inactivation of the enzyme by chemical modification techniques (48, 57).

The substitution of another metal for that present in the native state or the removal of any metal is the simplest chemical modification for a metalloenzyme. Marked changes in activity are usually observed in either case. Substitution of cadmium for zinc first demonstrated a difference in the esterase and peptidase activities of carboxypeptidase A (47). The activity of [(CPD)Cd] toward Bz-Gly-L-OPhe is increased, but that enzyme is virtually inactive toward Cbz-Gly-L-Phe.

Stopped-flow fluorescence studies of ES complexes provided a direct comparison of the peptide binding affinities of the zinc and cadmium enzymes and, simultaneously, an explanation for the different roles of metals in peptide and ester hydrolysis (48). Cadmium carboxypeptidase binds the peptide Dns-(Gly)$_3$-L-Phe as readily as does [(CPD)Zn] but catalyzes its hydrolysis at a rate that is reduced considerably (Figure 8). Initial rate studies of oligopeptides are in agreement with this observation. For all peptides examined, the catalytic rate constants of the cadmium enzyme are decreased markedly, but the association constants (1) (K_M^{-1} values) of the cadmium enzyme are identical to those of the zinc enzyme (48, 51, 57). However, in marked contrast, for all esters examined the catalytic rate constants of the cadmium enzyme are nearly the same as those of the zinc enzyme, but the association constants are decreased greatly.

Removal of the metal atom from carboxypeptidase A drastically reduces its activity toward both peptide and ester substrates (58, 59). It is important to determine if the loss in activity is attributable to weakened substrate binding or to reduced catalytic efficiency. Equilibrium binding techniques are precluded since the residual activity is sufficient to catalyze the conversion of the substrate into products before the measurements can be completed. Again, examination of the enzyme in the presence of substrates under stopped-flow conditions can resolve these questions. While the apoenzyme exhibits greatly reduced activities toward an exactly matched ester–peptide pair, the mechanisms for the

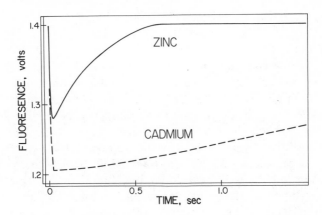

Biochemistry

Figure 8.   Tracings of stopped-flow fluorescence assays of the hydrolysis of Dns-(Gly)$_3$-L-Phe, 2.5 × 10^{-4}M, catalyzed by zinc and cadmium carboxypeptidase A, 5 × 10^{-5}M, at pH = 7.5 and 25°C in 0.03M Tris–1.0M NaCl (48). Enzyme tryptophans were excited at 285 nm and their emission was measured by means of a band-pass filter peaking at 360 nm.

loss of activity are different. Thus, the apoenzyme (CPD), although unable to catalyze the hydrolysis of peptide substrates, binds them to the same degree as does the zinc enzyme (Figure 9). However, when zinc is removed from the native enzyme, binding of the exact ester analog

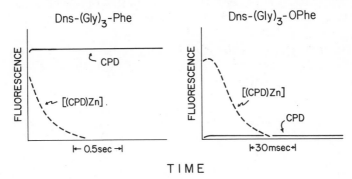

Enzyme Action

Figure 9.   Stopped-flow fluorescence measurements of Dns-(Gly)$_3$-L-Phe, 1 × 10^{-4}M, and Dns-(Gly)$_3$-L-OPhe, 1 × 10^{-4}M, both binding to zinc (dashed line) and apo- (solid line) carboxypeptidase A at pH = 7.5 and 25°C in 0.03M Tris–1.0M NaCl (48, 57). Concentration of the zinc and apoenzyme was 5 × 10^{-5}M for peptide and 2 × 10^{-5}M for ester hydrolysis. Enzyme tryptophans were excited at 285nm and dansyl emission was measured by means of a 430-nm cutoff filter.

decreases by orders of magnitude. Hence, the binding of peptides to metallocarboxypeptidase must differ from that of esters. Steady-state kinetic studies of the metalloenzymes are in agreement with the conclusions ($48, 51$). A comparison of the kinetic parameters for the zinc, cobalt, manganese, and cadmium enzyme-catalyzed hydrolysis of benzoyl-glycyl-glycyl-L-phenylalanine (Table IV) reveals a range of k_{cat} values from 6000 min^{-1} for the cobalt enzyme to 43 min^{-1} for the cadmium enzyme. The K_M^{-1} values, on the other hand, are essentially independent of the particular metal present. Thus, it would appear that the primary role of the metal in peptide hydrolysis is to function in the catalytic process and that it has little to do with peptide binding (48–53). On the other hand, metal substitution markedly alters the binding affinity of the exact ester analog, Bz-(Gly)$_2$-L-OPhe. The K_M^{-1} values decrease in the same order as do the k_{cat} values of peptide hydrolysis i.e., Co > Zn > Mn > Cd, the affinity of the ester for [(CPD)Co] being 30 times greater than that for [(CPD)Cd]. Metal substitution, however, has no significant effect on the catalytic rate constant of ester hydrolysis. These results indicate that the primary role of the metal in ester hydrolysis is to serve as a binding locus during catalysis (48).

Recent studies of Co(III) carboxypeptidases are also consistent with these proposals (60). The Co(III) enzyme is inactive toward both peptides and esters, but examination of ES complexes by stopped-flow fluorescence demonstrates that the peptide still binds to the modified enzyme while the ester does not. Since Co(III) complexes are "exchange inert," these results suggest that an innersphere complex is formed between the ester and the metal atom during binding but not for the peptide.

Inhibitors which are thought to bind to the active site metal atom also differentiate ester from peptide binding. The effect of an inhibitor on the fluorescent characteristics of an ES complex readily identifies the

Table IV. Metallocarboxypeptidase-Catalyzed Hydrolysis of Bz-(Gly)$_2$-L-Phe and Bz-(Gly)$_2$-L-OPhe[a]

| | *Bz-(Gly)$_2$-L-Phe* | | *Bz-(Gly)$_2$-L-OPhe* | |
Metal	k_{cat} *(min^{-1})*	$10^{-3}K_M^{-1}$ *(M^{-1})*	$10^{-4}k_{cat}$ *(min^{-1})*	K_M^{-1} *(M^{-1})*
Cobalt	6000	1.5	3.9	3300
Zinc	1200	1.0	3.0	3000
Manganese	230	2.8	3.6	660
Cadmium	41	1.3	3.4	120

[a] Assays performed at 25°C, pH = 7.5, 1.0M NaCl, and a buffer concentration of 0.05M Tris for peptide hydrolysis and $10^{-4}M$ Tris for ester hydrolysis (48).

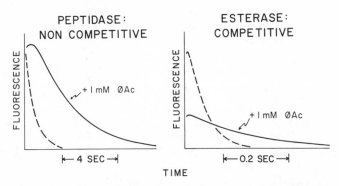

PEPTIDASE:
NON COMPETITIVE

ESTERASE:
COMPETITIVE

Enzyme Action

Figure 10. Phenyl acetate inhibition of carboxypeptidase A-catalyzed hydrolysis of peptide, Dns-(Gly)₃-L-Phe, 1 × 10⁻⁴M, and the ester, Dns-(Gly)₃-L-OPhe, 4 × 10⁻⁵M (57). Enzyme concentrations were 5 × 10⁻⁶ and 4 × 10⁻⁵M for ester and peptide hydrolysis, respectively (48, 57). Assays were performed in the absence (dashed line) and presence (solid line) of 1 × 10⁻³M phenyl acetate, øAc, at 25°C and pH = 6.5, 0.03M Mes–1.0M NaCl. Excitation was at 285 nm and dansyl emission, above 430 nm, was observed.

mode of inhibition (56, 57). The behavior of phenyl acetate toward the native enzyme · Dns-(Gly)₃-L-OPhe complex at pH = 6.5, for example, is characteristic of competitive inhibition (Figure 10). In the presence of phenyl acetate, 1 mmol, the initial peak height is reduced, but the area under the curve remains unchanged. The K_I was determined to be 3.2 × 10⁻⁴M. However, when phenyl acetate acts on the ES complex of the exact peptide analog Dns-(Gly)₃-L-Phe (Figure 10), the area under the tracing increases greatly, consistent with a markedly reduced rate of hydrolysis. Since the inhibitor does not decrease the initial peak height, substrate binding has not been decreased. This is characteristic of non-competitive inhibition (56). The K_I value, 3.3 × 10⁻⁴M, is the same as that obtained for ester hydrolysis. Similar results have been obtained by steady-state means with a number of such inhibitors acting on matched peptide and ester pairs (48, 57). In each case the K_I values are identical (Table V), suggesting identical enzyme–inhibitor complexes. However, since bound inhibitor allows the peptide but not the ester to bind to the enzyme, the binding of these substrates to the locus at which the inhibitor interacts with the enzyme must differ.

These results can be summarized in terms of the mechanisms shown in Figure 11 for peptide and ester hydrolysis. Assuming the peptide binds to the enzyme in the manner shown above, the exact ester analog must bind in a different manner, particularly with regard to the active site metal atom. The placement of the peptide carbonyl group on the

Table V.   Carboxylic Acid Inhibitors of the Carboxypeptidase A-Catalyzed Hydrolysis of Esters and Peptides[a]

Substrate	Inhibitor	$K_I \times 10^4$ (M)	Type of Inhibition
Dns-(Gly)₃-L-Phe	phenyl acetate	3.3	noncompetitive
Dns-(Gly)₃-L-OPhe	phenyl acetate	3.2	competitive
Cbz-(Gly)₂-L-Phe	indole 3-acetate	1.7	noncompetitive
Bz-(Gly)₂-L-OLeu	indole 3-acetate	1.6	competitive
Bz-(Gly)₂-L-Phe	β-phenyl propionate	1.2	noncompetitive
Bz-(Gly)₂-L-OPhe	β-phenyl propionate	1.2	competitive

[a] Assays performed at 25°C, pH = 7.5, 1.0M NaCl, 0.5M Tris except for the phenyl acetate study, where the conditions were pH = 6.5, 1.0M NaCl, 0.03M 2-(N-morpholino)-ethanesulfonic acid (Mes) buffer (48).

Biochemistry

metal is consistent with the metal functioning primarily in the catalytic step of peptide hydrolysis. The placement of the ester carboxyl group on the metal is consistent with the metal functioning in the binding step of ester hydrolysis and with chemical modification studies, which rule out the arginyl residue as essential for ester binding (61). Inhibitors that bind to the metal through a carboxyl group could then disrupt the metal interaction with peptide carbonyl and the ester carboxyl group, resulting in noncompetitive and competitive inhibition, respectively, as is observed. These results clearly demonstrate that there must be different mechanisms for the hydrolysis of esters and peptides which differ only in the susceptible bond.

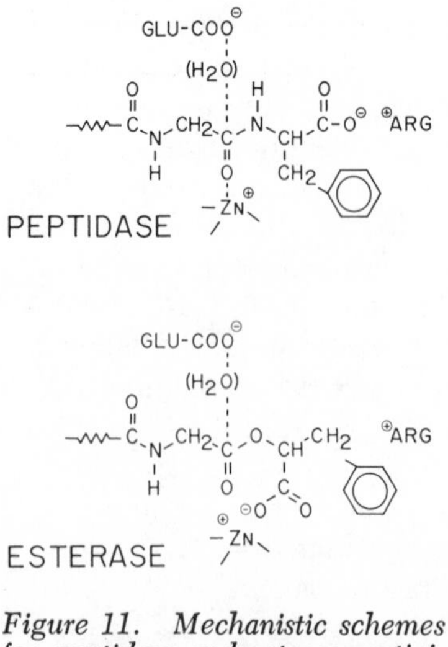

Figure 11.   Mechanistic schemes for peptidase and esterase activities of carboxypeptidase A

Summary

Definitive knowledge that zinc is indispensable to living matter has emerged only in the last two generations. In succession, the biological effects of zinc were viewed as: mostly harmful, questionable, and now essential. However, it is now well established that zinc is essential for the growth and development of all living forms. Zinc is now known to participate in a variety of processes, including protein, nucleic acid, carbohydrate, and lipid metabolism. It is present throughout all phyla and is essential for the function and/or structure of at least one in each of the six categories of enzymes designated by the Commission on Enzyme Nomenclature of the International Union of Biochemistry; among them the alcohol dehydrogenases, aldolases, alkaline phosphatases, carboxy-peptidases, leucine amino peptidases, neutral bacterial proteases, DNA- and RNA polymerases, an isomerase, transphosphorylase, and aspartate transcarbamylase. Identification of zinc metalloenzymes is a relatively recent occurrence. Two decades ago only three or four zinc enzymes were known while more than 90 have been identified today, and the number is steadily increasing. Detailed studies of the structure and function of many of these enzymes have become central to present under-standing of enzymatic catalysis. Mechanistic studies of carboxypeptidase A have defined the role of the metal in catalysis, the metal functions in the catalytic step of peptide hydrolysis and in the binding step of ester hydrolysis.

While much is now known about the mode of action of zinc metallo-enzymes in vitro, understanding the role of zinc in vivo is yet to be accomplished. In an effort to bridge this gap, studies have been carried out with the eukaryotic organism, *E. gracilis,* to examine the events accompanying zinc deficiency. Each step in the growth cycle of this alga requires zinc, but it is not yet possible to account for these results in terms of specific zinc metalloenzymes. Some of the difficulties which preclude such assignments are the possible existence of the same activity, being present in zinc-dependent and zinc-independent enzymes and the ability of other metals to substitute for zinc with minimal impairment of function. Means should be sought to determine the activity and metal content of such enzymes in the cellular environment in order to relate composition and biologic function.

Literature Cited

1. Raulin, J., *Ann. Sci. Nat., Bot. Biol. Veg.* (1869) **11**, 93.
2. Lechartier, G., Bellamy, F., *Compt. Rend. Acad. Sci.* (1877) **84**, 687.
3. Raoult, F., Breton, H., *Compt. Rend. Acad. Sci.* (1877) **85**, 40.
4. Bertrand, G., Bhattacherjee, R. C., *Compt. Rend. Acad. Sci.* (1934) **198**, 1823.

5. Moynahan, E. J., *Lancet* (1974) **2**, 399.
6. Keilin, D., Mann, T., *Biochem. J.* (1940) **34**, 1163.
7. Vallee, B. L., Neurath, H., *J. Biol. Chem.* (1955) **217**, 253.
8. Preyer, W. T., "De Haemoglobino Observationes et Experimenta," p. 27, M. Cohen and Son, Bonn, 1866.
9. Harless, E., Müller's *Arch. Anat. Physiol., Physiol. Abt.* (1847) 148.
10. Vallee, B. L., *Trends Biochem. Sci.* (1976) **1**, 88.
11. Price, C. A., Vallee, B. L., *Plant Physiol.* (1962) **37**,·428.
12. Falchuk, K. H., Fawcett, D. W., Vallee, B. L., *J. Cell. Sci.* (1975) **17**, 57.
13. Falchuk, K. H., Krishan, A., Vallee, B. L., *Biochemistry* (1975) **14**, 3439.
14. Wacker, W. E. C., *Biochemistry* (1962) **1**, 859.
15. Price, C. A., "Zinc Metabolism," A. S. Prasad, Ed., p. 69, C. C. Thomas, Springfield, Illinois, 1966.
16. Howard, A., Pelc, S. R., "Symposium on Chromosome Breakage" (Supplement to "Heredity," Vol. 6), Charles C. Thomas, Springfield, Illinois, 1953.
17. Baltimore, D., *Nature* (1970) **22**, 1209.
18. Temin, H. M., Mizutani, S., *Nature* (1970) **226**, 1211.
19. Vallee, B. L., Fluharty, R. G., Gibson, J. G., 2nd, *Acta Unio Int. Cancrum* (1949) **6**, 869.
20. Gibson, J. G., 2nd, Vallee, B. L., Fluharty, R. G., Nelson, J. E., *Acto Unio Int. Cancrum* (1950) **6**, 1102.
21. Auld, D. S., Kawaguchi, H., Livingston, D. M., Vallee, B. L., *Biochem. Biophys. Res. Commun.* (1974) **57**, 967.
22. Auld, D. S., Kawaguchi, H., Livingston, D. M., Vallee, B. L., *Proc. Nat. Acad. Sci. U.S.A.* (1974) **71**, 2091.
23. Kawaguchi, H., Vallee, B. L., *Anal. Chem.* (1975) **47**, 1029.
24. Kawaguchi, H., Auld, D. S., *Clin. Chem.* (1975) **21**, 591.
25. Auld, D. S., Kawaguchi, A., Livingston, D. M., Vallee, B. L., *Biochem. Biophys. Res. Commun.* (1975) **62**, 296.
26. Auld, D. S., Atsuya, I., Campino, C., Valenzuela, P., *Biochem. Biophys. Res. Commun.* (1976) **69**, 548.
27. Falchuk, K. H., Mazus, B., Ulpino, L., Vallee, B. L., *Biochemistry* (1976) **15**, 4468.
28. Falchuk, K. H., Ulpino, L., Mazus, B., Vallee, B. L., *Biochem. Biophys. Res. Commun.* (1977) **74**, 1206.
29. McLennan, A. G., Keir, H. M., *Biochem. J.* (1975) **151**, 239.
30. Kotsiopoulos, S., Mohr, S. C., *Biochem. Biophys. Res. Commun.* (1975) **67**, 979.
31. Psorske, L. M., Auld, D. S., Cohn, M., *Fed. Proc., Fed. Am. Soc. Exp. Biol.* (1977) **36**, 706.
32. Scrutton, M. C., Wu, C. W., Goldthwait, D. A., *Proc. Nat. Acad. Sci. U.S.A.* (1971) **68**, 2497.
33. Slater, J. P., Mildvan, A. S., Loeb, L. A., *Biochem. Biophys. Res. Commun.* (1971) **44**, 37.
34. Springgate, C. F., Mildvan, A. S., Abramson, R., Engle, J. L., Loeb, L. A., *J. Biol. Chem.* (1973) **248**, 5987.
35. Coleman, J. E., *Biochem. Biophys. Res. Commun.* (1974) **60**, 641.
36. Vallee, B. L., "Biological Aspects of Inorganic Chemistry," D. Dolphin, Ed., p. 37, John Wiley & Sons, New York, 1977.
37. Vallee, B. L., Williams, R. J. P., *Proc. Nat. Acad. Sci. U.S.A.* (1968) **59**, 498.
38. Vallee, B. L., Latt, S. A., "Structure-Function Relationships of Proteolytic Enzymes," P. Desnuelle, H. Neurath, M. Ottesen, Eds., p. 144, Academic, New York, 1970.
39. Latt, S. A., Vallee, B. L., *Biochemistry* (1971) **10**, 4263.
40. Kaden, T. A., Holmquist, B., Vallee, B. L., *Inorg. Chem.* (1974) **13**, 2585.

41. Kennedy, F. S., Hill, H. A. O., Kaden, T. A., Vallee, B. L., *Biochem. Biophys. Res. Commun.* (1972) **48**, 1533.
42. Riordan, J. F., Vallee, B. L., *Adv. Exp. Med. Biol.* (1974) **48**, 33.
43. Quiocho, F. A., Lipscomb, W. N., *Adv. Protein Chem.* (1971) **25**, 1.
44. Colman, P. M., Jansonius, J. N., Matthews, B. W., *J. Mol. Biol.* (1972) **70**, 701.
45. Brandén, C.-I., Jörnvall, H., Eklund, H., Furugren, B., "The Enzymes," P. D. Boyer, Ed., p. 104, Academic, New York, 1975.
46. Felber, J.-P., Coombs, T. L., Vallee, B. L., *Biochemistry* (1962) **1**, 231.
47. Coleman, J. E., Vallee, B. L., *J. Biol. Chem.* (1960) **235**, 390.
48. Auld, D. S., Holmquist, B., *Biochemistry* (1974) **13**, 4355.
49. Auld, D. S., Holmquist, B., *Fed. Proc., Fed. Am. Soc. Exp. Biol.* (1972) **31**, 1230.
50. Auld, D. S., Holmquist, B., *IXth International Congress of Biochemistry, Stockholm*, 1973, p. 65.
51. Auld, D. S., Vallee, B. L., *Biochemistry* (1970) **9**, 602.
52. Auld, D. S., Vallee, B. L., *Biochemistry* (1970) **9**, 4352.
53. Auld, D. S., Vallee, B. L., *Biochemistry* (1971) **10**, 2892.
54. Latt, S. A., Auld, D. S., Vallee, B. L., *Proc. Nat. Acad. Sci. U.S.A.* (1970) **67**, 1383.
55. Latt, S. A., Auld, D. S., Vallee, B. L., *Biochemistry* (1972) **11**, 3015.
56. Auld, D. S., Latt, S. A., Vallee, B. L., *Biochemistry* (1972) **11**, 4994.
57. Auld, D. S., "Enzyme Action," *in* "Biorganic Chemistry," E. E. Van Tamlen, Ed., Vol. I, p. 1, Academic, New York, 1977.
58. Vallee, B. L., Riordan, J. F., Coleman, J. E., *Proc. Nat. Acad. Sci. U.S.A.* (1963) **49**, 109.
59. Vallee, B. L., Riordan, J. F., Auld, D. S., Latt, S. A., *Philos. Trans. R. Soc. London, Ser. B* (1970) **257**, 215.
60. Van Wart, H. E., Vallee, B. L., *Biochem. Biophys. Res. Commun.* (1977) **75**, 732.
61. Riordan, J. F., *Biochemistry* (1973) **12**, 3915.
62. Vallee, B. L., *Adv. Protein Chem.* (1955) **10**, 317.

RECEIVED March 20, 1978. This work was supported by Grant-in-Aid Gm-15003 from the National Institutes of Health of the Department of Health, Education, and Welfare.

10

A Modified Standard Addition Method for Determining Cadmium, Lead, Copper, and Iron in Sea Water Derived Samples by Atomic Absorption Spectroscopy

C. P. WEISEL and J. L. FASCHING

Department of Chemistry, University of Rhode Island, Kingston, RI 02881

S. R. PIOTROWICZ[1] and R. A. DUCE

Graduate School of Oceanography, University of Rhode Island, Kingston, RI 02881

Flameless atomic absorption spectroscopy using the heated graphite furnace is a sensitive method for analyzing environmental samples for trace metals. High salt concentrations cause interference problems that are not totally correctable by optimizing furnace conditions and/or using background correctors. We determined that samples with identical ratios of major cations have trace metal absorbances directly related to their Na and trace metal concentrations. Equations and curves based on the Na concentration, similar to standard addition curves, can be calculated to overcome the trace element interference problem. Concentrations of Pb, Cd, Cu, and Fe in sea water can be simply and accurately determined from the Na concentration, the sample absorbance vs. a pure standard, and the appropriate curve.

The development of the heated graphite furnace for flameless atomic absorption spectroscopy allows measurements to be made on extremely small volumes of samples with high sensitivity. For example, as few as 10^{11} atoms of copper can be measured accurately. For these reasons, this

[1] Present address: Energy Resources Company, Inc., 185 Alewife Brook Parkway, Cambridge, MA 02138

0-8412-0416-0/79/333-172-134$05.00/0

technique has been used extensively in environmental and analytical chemistry, but a number of serious interferences are present in samples with high saline content such as sea water (*1, 2*). These complex matrix problems persist even with the use of background correctors which compensate for light scattering absorption and emission from particles over a continuum of wavelengths (*3*). A number of different approaches have been attempted to alleviate this problem, including additions of acids (*4, 5*) and other chemicals (*6*) to stabilize the element being determined or to volatilize interfering species and optimization of time and temperature settings (*7*). Investigations of the complex reactions occurring in the furnace and isolations of the ones of interest are being carried out (*5, 8*). These approaches have met with success for particular matrix systems and elements, but a generalized procedure has not been developed. This has resulted in the use of traditional methods such as standard addition and matrix matching of samples and standards (*6, 9*). Chemical processing of the samples to remove the interfering salts by solvent

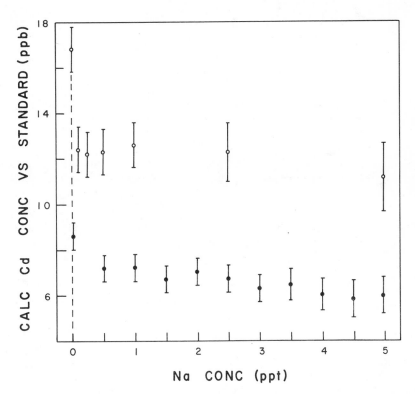

Figure 1. The calculated concentration of cadmium for two different level as a function of sodium concentration in the matrix

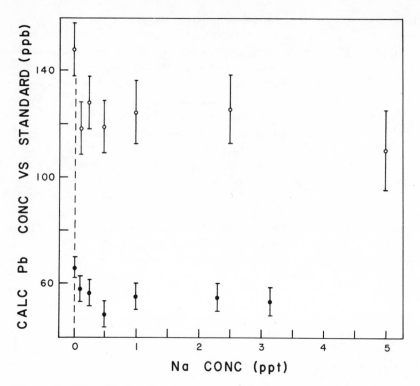

Figure 2.   The calculated concentration of lead for two different levels as a function of sodium concentration in the matrix

extraction (*10*), chelation techniques (*11*), and ion exchange chromatography have also been used to reduce interferences. These techniques, although generally successful, increase the number of samples to be analyzed, the preparation time, and the amount of sample required for a given analysis. The increased manipulation of samples also enhances the chances of their inadvertent contamination.

Experimental

To overcome these problems, we have developed an empirical interference correction technique. The technique was studied for the four elements in sea water; cadmium, copper, iron, and lead. A Perkin-Elmer 503 Atomic Absorption Spectrophotometer equipped with a heated graphite furnace (HGA2100) and a deuterium arc background corrector was used. The instrumental settings for each element are listed in Table I.

The method developed was based on analysis of 5-μL injections of standards prepared from comercially available 1000 ppm atomic absorption standards diluted with doubly deionized water, redistilled nitric

Table I.   Atomic Absorption Settings

Element	Wavelength (nm)	Drying Temp. (°C)	Time (sec)	Charring Temp. (°C)	Time (sec)	Atomization Temp. (°C)	Time (sec)
Cd	228.8	150	15	400	18	1700	7
Cu	324.7	150	20	1000	30	2500	6
Fe	302.1	150	20	1000	30	2500	6
Fe	372.0	150	20	1000	30	2500	6
Pb	283.3	150	15	550	18	2000	7

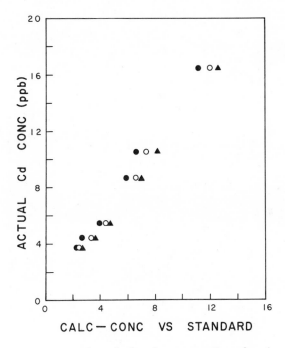

Figure 3.   The calculated concentration of various cadmium standards at three matrix levels of sodium is plotted vs. the actual cadmium concentration in the solution. (▲) 1.0 part per thousand sodium; (○) 3.0 parts per thousand sodium; (●) 5.0 parts per thousand sodium.

acid, and Suprapur hydrofluoric acid. Samples were prepared from Copen-
hagen Sea Water (IAPSO Standard Sea Water Service, Carlottenlund Slot,
DK-2920, Carlottenlund, Denmark) which was found to have very low
trace metal concentrations, nitric acid, hydrofluoric acid, and spiked to the
desired trace element concentration. The final acid concentration in all
solutions was 1.6N in nitric and 2.0N in hydrofluoric acid. The concentration of sea salt is expressed as a function of its sodium concentration and
ranged as high as 5.0 parts per thousand. Sodium was chosen since its
concentration is easily measurable and is proportional to the major ions in
sea water which are suspected as the cause of the interferences.

Results

For each of the elements examined, an initial suppression in signal
is observed with the introduction of sea salts. Figures 1 and 2 show this
effect for cadmium and lead respectively. The calculated concentrations,
based on the absorbance curve of de-ionized water-acid standards for
two series of spiked sample solutions for each element, are plotted against
the concentration of sea salt present in solution (expressed as sodium).

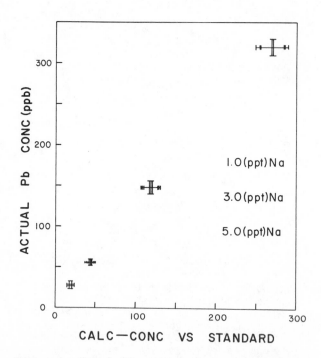

*Figure 4. Calculated concentration of various lead
standards at three matrix levels of sodium is plotted vs.
the actual lead concentration in the solution.*

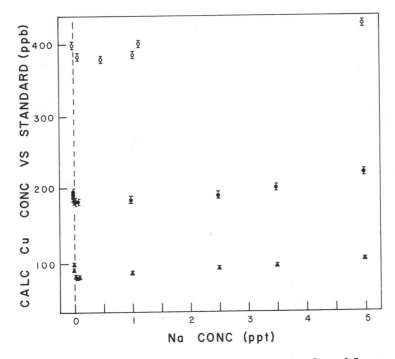

*Figure 5. The calculated concentration of copper for three different
levels as a function of sodium concentration in the matrix*

In each figure two different levels of the metal are depicted, with the
error bars illustrating one standard deviation. Both elements exhibit an
initial rapid decrease in calculated concentration up to a sea salt concentration of 0.5 parts per thousand sodium. This depression amounts to
~ 40% and ~ 25% of the actual concentration present for cadmium and
lead respectively. The initial decrease is followed by a leveling region,
which seems to continue to decrease slightly for cadmium while forming
a flat plateau for lead. The differences between the two metals in this
region are more clearly illustrated in Figures 3 and 4. The actual concentration of metal present is plotted against the value calculated from
the pure standard curves. Data for three sea salt concentrations, 1.0, 3.0,
and 5.0 parts per thousand sodium, are plotted. The coincidental points
in Figure 4 for lead indicate that the plateau region is flat and that the
observed signal does not change with varying salt content in this range.
Figure 3, however, shows the trend of decreasing signal with increasing
salt content for the indicated salt range and cadmium concentrations.

The variation of copper's signal with sea salt is somewhat different
from cadmium or lead as seen for the three metal concentrations shown

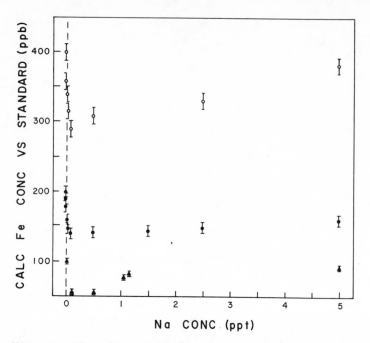

Figure 6.   The calculated concentration of iron (302-nm analyte line) for three different levels as a function of sodium concentration in the matrix

in Figure 5. An initial depression is again observed, but here the calculated concentration is only ~ 5% less than the actual concentration. In addition, instead of a plateau or decreasing region above 1.0 parts per thousand sea salt as sodium, a dramatic rise is found. The concentration calculated for the 5.0 parts per thousand solutions is slightly larger than that actually present.

Figures 6 and 7 give the variation in calculated iron concentration as a function of sodium in the sea salt when measured at 302 and 372 nm. The iron response to increasing salt concentration depends upon the wavelength used for analysis. Each plot has an initial suppression of the signal for the three calculated concentration levels shown for 302-nm wavelength and the one concentration for 372-nm wavelength. The depression is ~ 30% below the actual level present. Both wavelengths yield the same percent depression which supports our original contention that the observed decrease in the signal for each element is attributable to an interaction between the metal and sea salt, and the magnitude of the depression is a function of the metal being examined and the furnace parameter values.

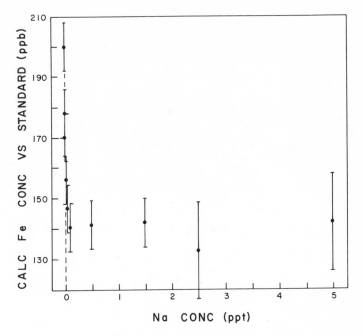

Figure 7. The calculated concentration of iron (372-nm analyte line) for one level as a function of sodium concentration in the matrix

Discussion

The cause of the rise of the plots of copper and iron (302 nm), the plateau regions of the plots of lead and iron (372 nm), and the decreasing region of the plots of cadmium above 1.0 part per thousand sea salt as sodium has a wavelength dependence rather than an elemental dependence. The above observations infer that at least two observable processes are occurring. The first phenomena is a decline in signal which is nearly complete for all elements with the addition of 1.0 part per thousand sodium–sea salt and whose percent change in signal is element dependent. The magnitude of the second effect is related to wavelength, sea salt concentration, and analyte concentration. One highly speculative explanation for the second effect can be based on the deuterium arc's energy profile vs. wavelength and Rayleigh scattering inverse dependence on the fourth power of the wavelength (Figures 8 and 9). As the wavelengths used vary from 228 nm for cadmium to 372 nm for iron, it is observed that the energy of the deuterium arc goes from a maximum to its lowest value. Both cadmium and lead are in a relatively high intensity region of the arc's spectrum. This implies that the analyte line's and the

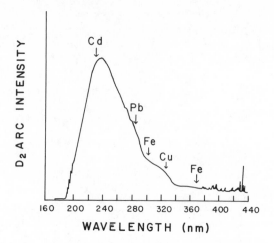

Figure 8. The deuterium arc intensity as a function of wavelength of the analyte lines

deuterium arc's energies can be balanced at a high intensity level and in a very stable region of the arc. More correction of light scattering is therefore available for the elements measured at the shorter wavelengths than for copper and iron. This is supported by the trend seen for the four lowest wavelengths. Cadmium at 228 nm has a slight decrease in the signal with increasing sea salt, and lead at 283 nm has a flat region, indicating that there is complete compensation of light scattering up to

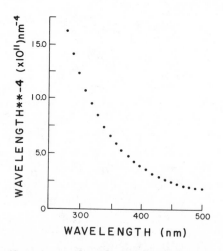

Figure 9. The relative change in the Rayleigh scattering as a function of wavelength

the 5.0 parts per thousand sea salt as sodium. The iron measured at 302 nm, in a higher intensity portion of the arc's energy than copper at 324 nm, shows a rise in signal but not as rapidly nor to as great an extent as that found for copper at the 5.0 parts per thousand sodium–sea salt level.

Further support for this speculative explanation is the decrease in sodium concentration necessary to cause the crossover point (a declining-to-rising signal) as the copper and iron (302 nm) concentrations in the samples increase. Furthermore, as the concentration and therefore the absorption by the element increases, the relative amount of arc balanced by the element and therefore that usable for compensation of light scattering decreases. Thus, visible matrix interferences occur at lower sea salt levels. This is also a reason why the amount of corrected light scattering is different between blanks and samples containing equivalent amounts of sea salts, and therefore the total signal from light scattering is not subtracted as part of a blank correction. Other possible phenomena exist that can also explain these observations.

Based solely on the above arguments it would be predicted that when iron is measured at 372 nm, which is on the minimum portion of the arc's energy, a rise in signal should occur similar to or larger than that of copper. This, however, does not occur. It may be because the amount of observed light scattering also is attributable to the efficiency of the particles present to scatter light of a particular wavelength. Rayleigh scattering by particles is inversely proportional to the fourth power of the wavelength and, as shown in Figure 9, drops very rapidly from the copper wavelength used to the iron wavelength of 372 nm. Since the same solution and temperature settings were used, one can assume a similar size particle range is present. Based on wavelength differences alone, 1.7 times more scattering should occur for copper than for iron.

The data and arguments presented indicate that the presence of sea salts alters the atomic absorption signal of these four elements from their response in de-ionized water–acid standards in a consistent manner for each particular element and sea salt concentration. This observation can be used to develop a modified standard addition technique. If a series of curves can be prepared that contain the range of metal and sea salt concentration expected in the samples, correction factors between actual and observed concentrations based on pure water–acid standards can be determined. This modified standard addition technique is illustrated for two of the elements discussed previously. For lead the actual concentration is plotted vs. the calculated concentration for a sea salt range of 1.0–5.0 parts per thousand sodium in Figure 4. This plot was prepared from solutions of known concentrations in a sea water medium. For a sample of unknown lead concentration, within the specified range,

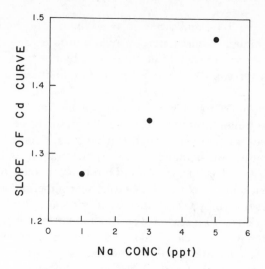

Figure 10. The change in the slope of the cadmium standardization curves for the various level of sodium in the matrix

the actual concentration can be obtained from a calculated one based on acid standards and sodium concentration by using this plot. If preferred, linear regression can be done to obtain this relationship in functional form. For this data, the following equation was obtained:

$$(\text{Pb}_{\text{actual}}) = 1.28 \, (\text{Pb}_{\text{calculated}})$$

The second illustration, using cadmium, has a more complicated relationship among the actual and calculated concentrations and the sea salt concentrations in the 1.0–5.0 parts per thousand range (Figure 3). Here the slope of the actual vs. calculated concentration line decreases with increasing sea salt concentration. The change of slope with sea salt was found to be linear (Figure 10). It is, therefore, graphically possible to estimate the actual concentration in an unknown from a plot such as Figure 3, using the calculated concentration and the sea salt sodium concentration. The relationship also can be expressed mathematically using the relationship among the slope and sodium concentration (Figure 10) and the actual and the calculated concentrations (Figure 3). The cadmium data is fitted by the equation:

$$(\text{Cd}_{\text{actual}}) = [\,(\text{Na parts per thousand}) \cdot 5 \times 10^{-2} + 1.21\,] \, (\text{Cd}_{\text{calculated}})$$

Conclusions

The additional preparation of a large number of samples for standard addition or matrix matching of each sample and standard can be replaced by the modified standard addition technique presented above. The preparations of curves in the desired sea salt range and metal concentrations need only be prepared once. This approach is applicable to any salt range whose effect on the signal is systematic and can be described by one or a series of plots or equations. Samples are compared with pure water–acid standards and sodium measurements, which are often already determined as part of an environmental analysis, to obtain the concentration in the unknowns. Care must be taken throughout to use the same matrix type in preparing the curves as that used in the samples, to use the same instrumental settings, and to have proper alignment of the background correction beam. This method also is readily adaptable to automated techniques and to calculations on programable calculators since simple equations are produced, relating the observed and actual concentrations.

Acknowledgment

The authors would like to thank Gerald Hoffman for access to unpublished data and for many helpful discussions and John Maney for his insights into this problem. We also wish to thank Andrew T. Zander for his helpful comments and criticisms. This work was supported by the National Science Foundation Office for the International Decade of Ocean Exploration under NSF/IDOE Grant Number OCE76-16883.

Literature Cited

1. Fuller, C. W., *Analyst* (1976) **101**, 798–802.
2. Pritchard, M. W., Reeves, R. D., *Anal. Chim. Acta.* (1976) **82**, 103–111.
3. Zander, A. T., *American Lab.* (1976) **8**, 11–22.
4. Regan, J. G. T., Warren, J., *Analyst* (1976) **101**, 220–221.
5. Yasuda, S., Kakiyama, H., *Anal. Chim. Acta.* (1977) **89**, 369–376.
6. Walsh, P. R., Fasching, J. L., Duce, R. A., *Anal. Chem.* (1976) **48**, 1014–1016.
7. Hendrix-Jongerius, C., DeGalan, L., *Anal. Chim. Acta.* (1976) **87**, 259–271.
8. French, W., Cedergen, A., *Anal. Chim. Acta.* (1977) **88**, 55–67.
9. Begnoche, B. C., Risby, T. H., *Anal. Chem.* (1975) **47**, 1041–1045.
10. Stolzberg, R. J., "Analytical Methods in Oceanography," T. R. P. Gibb, Jr., Ed., 30–43, ACS ADV. IN CHEM. **147**, Washington, DC, 1975.
11. Wright, F. C., Riner, J. C., *Atom. Absorb. Newsl.* (1975) **14**, 103–104.

RECEIVED January 16, 1978.

11

Sources of Environmentally Important Metals in the Atmosphere

ROBERT E. LEE, JR. and F. VANDIVER DUFFIELD

U.S. Environmental Protection Agency, Health Effects Research Laboratory, Research Triangle Park, NC 27711

Biological, chemical, and physical effects of airborne metals are a direct function of particle size, concentration, and composition. The major parameter governing the significance of natural and anthropogenic emissions of environmentally important metals is particle size. Metals associated with fine particulates are of concern; particles larger than about 3-μm aerodynamic equivalent diameter are minimally respirable, are ineffective in atmospheric interactions, and have a short air residence time. Seventeen environmentally important metals are identified: arsenic, beryllium, cadmium, chromium, copper, iron, mercury, magnesium, manganese, nickel, lead, antimony, selenium, tin, vanadium, and zinc. This report reviews the major sources of these metals with emphasis on fine particulate emissions.

It may be safe to say that the interest of environmental scientists in airborne metals closely parallels our ability to measure these components. Before the advent of atomic absorption spectroscopy, the metal content of environmental samples was analyzed predominantly by wet or classical chemical methods and by optical emission spectroscopy in the larger analytical laboratories. Since the introduction of atomic absorption techniques in the late 1950s and the increased application of x-ray fluorescence analysis, airborne metals have been more easily and more accurately characterized at trace levels than previously possible by the older techniques. These analytical methods along with other modern techniques such as spark source mass spectrometry and activation analysis

0-8412-0416-0/79/33-172-146$06.50/0

have enabled environmental scientists to better understand the relationships among airborne levels of metals, their sources, and their effect on human health.

Although these tools can help characterize the chemical composition of airborne particles in detail and with sensitivity never before possible, it is easy to generate an enormous amount of data which can literally swamp the environmental scientist. He is now faced with a new problem which concerns the selective analysis of airborne particulate matter that can provide the most useful information within the constraints of his resources. The question can be stated more simply: what are the environmentally important trace metals?

At least 11 metals are biologically toxic, as shown in Table I along with their threshold limit values (1, 2, 3): arsenic, beryllium, bismuth, cadmium, mercury, nickel, lead, antimony, selenium, tin, and vanadium. For these and other metals, the biological toxicity for long-term, low-level exposures is not known. Other metals such as manganese and iron are

Table I. Biologically Toxic Metals

Metal	Threshold Limit Value (1) (mg/m^3)	Effects of Excessive Exposures (2, 3)
As	0.5	Produces dermatitis, bronchitis, irritation of upper respiratory tract; may cause skin cancer
Be	0.002	Produces severe pulmonary reaction (berylliosis) and body-wide morbidity; carcinogenic to rats
Cd	0.05	Produces lung damage; may contribute to hypertension and cardiovascular disease; may be carcinogenic
Hg	0.05	Damages the brain and central nervous system; toxic to the kidney
Ni	1	Carcinogenic (as nickel carbonyl); produces dermatitis
Pb	0.15	Cellular poison; damages the liver, kidney, brain, and central nervous system; produces mental retardation
Sb	0.5	Shortens lives of rats and mice
Se	0.2	Irritates eyes, nose, throat, and respiratory tract; damages kidney and liver; acute poison; may be a carcinogen
Sn	2	Can produce neurologic damage and pneumoconiosis; accumulates in heart, intestine, and lung with age
V	0.05	Can produce bronchitis, emphysema, and pulmonary edema (V_2O_5); interferes with cholesterol synthesis

Table II. Environmentally Important Metals

As Be Cd Hg Ni Pb Sb Se Sn V	} Recognized as biologically toxic

Cu Fe Mg Mn Cr Ti Zn	} Long-term exposure effects not known / Catalyze secondary reactions / Damage materials / Absorb gaseous pollutants

important because of their role in catalyzing secondary gas-phase reactions in air, e.g., $SO_2 \rightarrow SO_4^=$, because of their effects on materials and because they can serve as adsorption sites for gaseous pollutants that can lead to enhanced contact toxicity. From these considerations, it is possible to compile a list (Table II) of environmentally important metals, albeit not necessarily one with which all would agree. The final consideration pertains to their presence or absence in ambient air. All of the metals given in Table II have been found in air; although the toxic properties of bismuth are recognized, the airborne concentration of this metal is too low at this time to present an environmental problem and therefore is not included in our listing ($4, 5, 6, 7, 8$).

Concentration and Particle Size Considerations

In the United States it is estimated that more than 15 million tons of particulate matter from anthropogenic sources are emitted into the air each year (9). Natural sources of particulate emissions from windborne dust, volcanic eruptions, and sea spray can contribute more than 10 times this amount. These estimates do not take into consideration the quantity of particles formed through photochemical and other atmospheric reactions; however, gas-to-particle reactions are not likely to generate new metal-containing particles.

It is difficult to ascertain what proportion of the anthropogenic and natural particulate emissions are composed of the environmentally important metals. Available information suggests that approximately 5% of the anthropogenic emissions, representing over 750,000 tons per year in the

United States, are environmentally important metals (5). Of these, lead and iron make up the largest component. Similar estimates for natural emissions are not possible because of the lack of information, but it is likely that magnesium from sea spray and iron and titanium from wind-borne dust are the major environmentally important metals emitted.

Table III. Estimates of Total and Fine Particulate Airborne Emissions from Various Sources in the United States

Anthropogenic Sources	Total Emissions in 1973 (Ton/Year (9))	Estimated Fine Particulate (0.1–2μm d.) Emissions in 1975 (Ton/year (11))	% Fine Particulate Fraction
Fuel Combustion			
Electrical generation	3,569,763	1,201,000	33.6
Commercial/industrial	2,845,626	150,000	5.3
Residential	235,944	—	—
Total	[6,651,333]		
Industrial processing			
Metallurgical processing	1,613,242	751,000	46.6
Mineral products	3,855,212	2,041,700	53.0
Food/agricultural	633,752	16,000	2.4
Chemical manufacturing	270,908	—	—
Miscellaneous	648,791	—	—
Total	[7,051,905]		
Incineration			
Municipal	75,788	51,000	67.3
Residential	316,952	—	—
Commercial	175,050	—	—
Total	[567,790]		
Transportation			
Surface	969,651	900,000	92.8
Aircraft	161,795	30,000	18.5
Vessels	24,794	—	—
Total	[1,156,240]		
Natural Sources			
Fugitive dust	33×10^6 (12)	—	—
Fires	519,006	—	—
Volcanic eruptions (global)	$25-150 \times 10^6$ (13)	—	—
Sea spray	122×10^6 (14)[a]	340,000	0.3

[a] Calculation was derived assuming a U.S. coastline of 12×10^3 mile, an effected area of 200 mile inland, and an emission factor of 140 ton/day/100 mile².

As Winchester (10) and others (5, 6) have pointed out, particle size considerations are extremely important in assessing the environmental impact of airborne metals. From an environmental health viewpoint, estimates of total particulate emissions alone are not very meaningful. What is meaningful is information on the fine particle fraction that can be inhaled. We have attempted to compile this information in Table III for the major anthropogenic and natural emission sources. Although these data are estimates and not actual measurements, they are helpful in appraising the environmental hazard. Especially striking in Table III is the large fine particle fraction from known metal emission sources such as electrical power generation (coal), municipal incineration, metallurgical processing, mineral production, and surface transportation (predominantly automobiles). Most important, perhaps, is that an estimated 46.6% of the total particulate emissions from metallurgical processing (Table III), consisting largely of the environmentally important metals, is in a particle size range < 2 μm in diameter.

Natural Emission Sources

Despite the fact that natural sources of airborne particles in the United States emit many times the particulate loading of anthropogenic sources, it is likely that these particles, especially fugitive dust and sea spray, are

Table IV. Average Concentration of Environmentally

| Metal | Oceans (17) | | Volcanoes (18) | |
	Sea Water	Sea Salt	Lava Ash	Fumarole Deposits
As	0.003	0.09	—	—
Be	—	—	—	—
Cd	—	—	—	—
Cr	0.002	0.06	—	—
Cu	0.002	0.06	—	—
Fe	0.0034	0.105	9.8	6.7
Hg	0.00003	0.0009	—	—
Mg	1,272	39,200	—	—
Mn	0.008	0.025	1,990	1,200
Ni	0.005	0.15	—	—
Pb	—	—	—	—
Sb	0.0002	0.006	0.60	0.93
Se	0.004	0.12	1	265
Sn	NA	NA	NA	NA
Ti	0.005	0.15	—	—
V	0.005	0.15	270	210
Zn	0.015	0.45	491	200

ᵃ In ppm.

predominantly large particles (*15*) that rapidly settle out and are not environmentally important except in the vicinity of the emissions. Volcanic eruptions are capable of injecting large quantities of particulate matter many kilometers above the earth's surface. For example, Hobbs et al. (*16*) reported that the eruption of the St. Augustine volcano near Alaska in 1976 emitted particles at the rate of 10^5 kg per second, reaching a height of 3.8 km 1 hr after the initial eruption. Most of the mass was caused by aerosol < 40 μm in diameter. From these data, they deduced that the estimates of worldwide volcanic emissions of 25–150 $\times$ 10^6 ton/year (*17*) are too low.

A compilation of the average concentration of the environmentally important metals associated with natural emission sources is given in Table IV. Magnesium is an important metal emission from sea spray, soil, and crustal rock. Significant levels of chromium, iron, manganese, titanium, and vanadium also are found in dust particles. Zinc, vanadium, selenium, antimony, manganese, and iron are found in volcanic emissions that can be important globally because of the large quantities of particulate mass involved and the possible long residence time in the atmosphere associated with the injection height.

A recent report by Beauford et al. (*22*) indicates that vegetation can release submicrometer-sized particles that could be dispersed over large areas. Their experiments showed that approximately 9 kg of Zn and 5 g

Important Metals Associated with Natural Sources[a]

	Dust	
Soil (19, 20)	*Crustal Rock (17, 20)*	*Airborne Soil-Sized Particles Collected Over Oceans (21)*
5	1.8	—
6	4.5	—
0.5	0.2	—
200	100	119
20	55	168
38,000	50,000	52,000
0.1	0.08	—
6,300	20,900	—
850	950	1,338
40	75	96
10	15	588
—	0.2	—
0.01	0.05	—
NA	NA	NA
4,600	4,400	—
100	135	163
50	70	780

of Pb can be released from 1 km² of vegetation per year. They conclude that the earth's great vegetation-covered land mass contributes to the trace metal composition of the atmosphere.

Stationary Emission Sources

Stationary sources are the major contributors of most environmentally important metals in air. Flinn and Reimers (23) reported the annual airborne emissions of metals in the United States from stationary sources projected through 1983 based on production estimates and assuming no changes in processes or control technology. Their results, summarized in Table V, show projected increases in emissions from all environmentally important metals where data are available. Comparatively low concentrations (150–900 ton/year) of the highly toxic metals—beryllium, selenium, and mercury—were reported for the 1969–1971 period. Metals emitted in the highest concentrations are zinc (151,000 ton/year) and titanium (88,000 ton/year), although iron could be expected to exceed these levels.

Although data are incomplete, the major stationary emission sources for many of the environmentally important metals are given in Table VI. Smelters/metallurgical processes are predominant sources of airborne cadmium, chromium, copper, and manganese. Coal and oil combustion,

Table V.   Annual Air Emissions of Environmentally Important Metals from Stationary Sources (23)

	Tons of Pollutant		
Metal	1969–1971 Average	1978 Projection	1983 Projection
As	9,000	12,750	16,990
Be	150	200	260
Cd	3,000	4,090	5,050
Cr	12,000	14,980	17,800
Cu	13,500	20,680	24,070
Fe	unknown	unknown	unknown
Hg	800	1,160	1,560
Mg	unknown	unknown	unknown
Mn	19,000	25,840	31,720
Ni	6,000	10,940	17,500
Pb	9,300	11,840	14,370
Sb	350	460	550
Se	900	1,240	1,560
Sn	unknown	unknown	unknown
Ti	88,000 (24)	unknown	unknown
V	18,000	37,240	58,370
Zn	151,000	216,700	273,000

Table VI. Percent of Total Air Emissions of Some Environmentally Important Metals from Stationary Sources (24)

Metal	Smelter/ Metallurgical Processing	Coal Combustion	Oil Combustion	Incineration
Be	3	88	6	—
Cd	43	—	—	52
Cr	68	9	—	—
Cu	84	7	—	—
Mn	57	11	—	—
Ni	11	—	83	—
Pb	2	—	—	1
Ti	—	9	—	—
V	1	—	90	—

primarily from power generation plants, are important contributors of beryllium, nickel, and vanadium. Incineration is a major contributor of airborne cadmium; however, emissions from incinerators are largely dependent on the composition of the waste material burned. Other workers (25, 26) have reported emissions of arsenic, beryllium, cadmium, chromium, copper, iron, mercury, magnesium, manganese, nickel, lead, antimony, titanium, vanadium, and zinc from municipal incinerators over a wide concentration range.

Table VII presents a summary of the approximate concentration ranges of the environmentally important metals found in particulate matter collected past emission control devices of six important stationary sources. These data largely support the results in Table VI and show that smelters and metallurgical processing industries are major metal emitters. Iron and steel foundries discharge into the air particulate matter with a high content of chromium, copper, iron, magnesium, manganese, nickel, lead, and zinc. Brass and bronze refineries emit particles enriched in cadmium, copper, iron, magnesium, lead, and especially high levels of zinc (10,000–100,000 ppm). Surprisingly, particulates from lead smelters contain only moderate quantities of lead (0.1–100 ppm) and low quantities of cadmium, copper, magnesium, manganese, nickel, and vanadium. Coal-fired power plant particulate matter contain a wide range of metals in moderate concentrations, including chromium, copper, iron, magnesium, manganese, nickel, lead, titanium, and zinc. Moderate-to-high concentrations of arsenic, cadmium, chromium, copper, manganese, nickel, lead, antimony, and vanadium are found in particulate matter emitted from municipal incinerators and cement plants.

Particle size considerations are extremely important in assessing the environmental hazard of metal emissions from stationary sources. Table III indicates that more than 65% of the emissions from municipal incin-

Table VII.   Approximate Concentration Ranges[a] of Matter Collected from

Metal	Iron and Steel Foundries	Brass/Bronze Refineries	Lead Smelters
As	—	—	—
Be	—	—	—
Cd	—	100–1,000	0.01–1
Cr	10–100	—	—
Cu	10–100	100–1,000	0.01–0.1
Fe	1,000–100,000	100–10,000	—
Hg	—	—	—
Mg	100–1,000	10–1,000	1–10
Mn	10–1,000	—	0.01–1
Ni	10–1,000	—	0.01–1
Pb	10–1,000	100–10,000	0.1–100
Sb	—	—	—
Se	—	—	—
Sn	—	100–10,000	0.01–1
Ti	—	—	—
V	—	—	0.01–0.1
Zn	100–1,000	10,000–100,000	—

[a] In ppm.

erators are in the fine particle size range (0.1–2.0 μm diameter), suggesting that a large fraction of these metals are likely to be associated with fine (and presumably respirable) particles.

The particle size of metal emissions from smelters and metallurgical processes is likely to vary widely with the type of process and the emission controls used. Table III shows that about 45% of the particles emitted from metallurgical processes is in the fine particle range. Lee et al. (27) studied emissions from an electric arc furnace steel plant equipped with a baghouse control device and found that 57% of the total particulate matter was less than 1 μm in diameter. Little information on the particle size of specific metal emissions is available in the literature.

Emissions from coal-fired power plants have been the subject of intensive study by many workers (27, 28, 29, 30). However, oil-fired power plants also can be expected to emit trace metals. A comparison of the approximate metal content of typical coals and oils given in Table VIII indicates that oils contain at least 13 of the important metals with high (> 10 ppm) concentrations of vanadium, zinc, and nickel. Coal contains at least 14 of the environmentally important metals with very high (> 100 ppm) concentrations of iron, magnesium, titanium, and zinc.

Recently, information also has become available on trace metal emissions from an oil-fired power plant without a control device (33). Although variations can be expected in emissions from different sources,

11. LEE AND DUFFIELD *Sources of Metals* 155

**Environmentally Important Metals in Particulate
Various Sources (6)**

Coal-Fired Power Plants	Municipal Incinerators	Cement Plants
—	—	—
—	1–10	1–10
—	1,000–10,000	10–1,000
1–100	100–1,000	100–1,000
0.1–10	100–1,000	100–10,000
10–1,000	1,000–10,000	10,000–100,000
—	—	—
10–100	—	—
1–1,000	100–1,000	100–1,000
1–100	100–1,000	100–1,000
1–100	10,000–100,000	100–1,000
—	10–1,000	100–1,000
—	—	—
1–10	—	—
1–100	—	—
—	10–100	10–100
10–1,000	—	—

it is instructive to compare typical emissions from a coal-fired (in this case, equipped with an electrostatic precipitator control device) and an oil-fired power plant; Table IX presents the trace metal emissions normalized for 100 MW of power generation. The first difference pertains

Table VIII. Typical Concentrations of Environmentally Important Metals in Oils and Coals

Metal	Oils (ppm) (31)	Coals (ppm) (32)
As	0.01	14
Be	0.0004	—
Cd	0.01	2.5
Cr	0.3	14
Cu	0.14	15
Fe	—	19,200
Hg	10	0.2
Mg	—	500
Mn	0.1	49
Ni	10	21
Pb	0.3	35
Sb	—	—
Se	0.17	2
Sn	0.01	2
Ti	0.1	700
V	50	33
Zn	25	272

Table IX. Comparison of Environmentally Important Metals Emitted from a Typical Coal-Fired and an Oil-Fired Power Plant

Metal	Coal-Fired (27) ($\mu g/m^3$ per 100 MW)	Oil-Fired (33) ($\mu g/m^3$ per 100 MW)
As	—[a]	BDL
Be	—	—
Cd	0.1	BDL
Cr	0.7	44.2
Cu	BDL[b]	13.5
Fe	1340	90.4
Hg	—	5.8
Mg	—	109.6
Mn	BDL	7.7
Ni	1.3	371.3
Pb	1.4	5.8
Sb	6.8	BDL
Se	6.5	—
Sn	—	BDL
Ti	—	1.9
V	1.5	1203.8
Zn	0.7	3.9
Total particulate	8700	43,077 (34)

[a] Denotes not sought in the analysis.
[b] BDL denotes below detectable limit.

to the quantity of total particulate matter emitted, with the oil-fired generator discharging into the air almost five times the amount of the coal-fired plant. The predominant coal-fired plant metal emission is iron, with lesser quantities of cadmium, chromium, nickel, lead, selenium, antimony, vanadium, and zinc. In addition, other workers (28, 30) have reported that arsenic, mercury, magnesium, manganese, and titanium are also emitted from coal combustion. Similarly, a variety of environmentally important metals are emitted from oil-fired power plants, principally nickel and vanadium. Other metals discharged, shown in Table IX, include chromium, copper, iron, mercury, magnesium, manganese, lead, titanium, and zinc. A comparison of the data in Table IX shows that the trace metal emission concentrations from the oil-fired power plant are greater than those from a coal-fired power plant except for iron; some metal emissions such as vanadium and nickel greatly exceed those from coal-fired power plants.

The particle size aspect of particulate components emitted from power plants is especially intriguing. Table III indicates that the fine particulate fraction emitted from electrical generation (coal and oil combustion) is estimated to be greater than 33%, a much smaller fraction than from metallurgical processing ($> 45\%$) and from municipal incineration ($> 65\%$). Lee et al. (27) made direct particle size measurements on a coal-fired power plant equipped with an electrostatic precipitator and found that 14% of the particulate mass was less than 1 μm in diameter while 38% was between 1 and 5 μm in diameter. They also showed that many of the metals emitted were associated with fine particulates. For example, particles < 1 μm in diameter contained 35% of the V, 80% of the Cr, 50% of the Pb, 62% of the Sb, 20% of the Cd, 30% of the Se, and 44% of the Zn.

Lee and other workers (6, 33, 35) have also reported the apparent preferential concentration of metals in the emissions and fly ash from coal-fired power plants. Natusch et al. (35) found that lead, antimony, cadmium, selenium, arsenic, zinc, nickel, and chromium were environmentally important metals that increased markedly in concentration in fly ash as a function of decreasing particle size, results which largely support those reported by Lee et al. (27). The enrichment of metals in the fine particle fraction has been explained by volatilization (33, 35). Volatilized metals can remain gaseous, partially recondense, or completely recondense. Gaseous volatilization results in a high percentage of the metal contained in coal, e.g., Hg, being discharged through the stack. Partial or complete condensation will lead to an increase in the concentration of these elements in the fine particulate fraction of the fly ash. Condensation can occur by nucleation, resulting in small particles, or by deposition on available surfaces, also resulting in increased concentration in the small particles because of the greater specific surface area.

Mobile Emission Sources

The combustion product(s) of lead-containing additives to motor fuels dominate any discussion of trace metals emitted from mobile sources. Lead emissions from motor vehicles far exceed those from any other source: in 1970, 214,000 tons of lead were emitted from gasoline combustion compared with 16,000 tons from all other sources combined (36). Among environmentally important metals typically found in automotive exhaust, lead particulate emissions usually exceed those of other metals such as manganese, iron, copper, and nickel by at least one order of magnitude (*see* Table X). The exception to this case is zinc, emissions of which are highly variable, ranging from below detection limits to greater than 80 μg/mile (39, 40). These are closely associated with use of certain lubricants, motor oils, or fuel additives in which zinc is frequently found as a major constituent and, to a lesser extent, fuel composition (*see* Tables XI, XII, and XIII).

Table X. X-Ray Fluorescence[a] of Particulate Matter

	Pb (mg/mile)
Pre-catalyst cars (37)	
FTP (mean)[c]	40.5
HWFET	19.8
CFS	20.0
60 mph cruise (38)	NA
Catalyst cars (49-state standard) (39)	
50 mph cruise	
mean	0.028
median	0.015
range	0–0.325
Catalyst cars (California standard) (40)	
FTP	0.03
Dual catalyst (41)	
FTP	0.18
HWFET	0.03
CFS	0.07
Lean burn (251 engine) (40)	
FTP	6.69
HWFET	3.76
CFS	7.00
Stratified charge (40)	
FTP	ND
HWFET	ND
CFS	0.12
Rotary (40)	
FTP	0.40
Diesel (#2 National avg. fuel) (42)	
FTP	2.55
HWFET	2.50
CFS	2.00

[a] Except for pre-catalyst copper and iron (emission spectrometry).
[b] Legend: FTP = 1975 Federal test procedure, urban commuter route simulation, 7.51 miles, 19.9 mph average speed; HWFET = highway fuel economy test, two-lane highway simulation, 10.24 miles, 49.9 mph average speed; CFS = crowded

in Diluted Automotive Exhaust, Selected Metals[b]

Mn (mg/mile)	Cu (mg/mile)	Fe (mg/mile)	Ni (mg/mile)
NA	NA	NA	ND
NA	NA	NA	ND
NA	NA	NA	ND
NA	0.22	1.2	NA
NA	0.016	0.029	NA
NA	0	0.007	NA
NA	0–0.293	0–0.341	NA
ND	0.18	2.28	0.01
NA	0.12	0.56	4.12
NA	0.01	0.03	0.27
NA	0.02	0.07	0.50
0.05	0.06	1.17	ND
0.12	0.08	0.07	ND
0.07	0.08	0.15	ND
ND	ND	0.13	0.01
ND	ND	0.23	0.01
0.01	ND	0.05	ND
0.09	0.04	0.70	0.06
1.46	1.56	1.56	NA
1.45	2.03	0.12	NA
1.19	1.54	0.08	NA

freeway simulation, 13.59 miles, 35.0 mph average speed; NA = not available; ND = not detected.

[c] All values reported are means unless otherwise specified.

Table XI.   Selected #2 Distillate

Source/ Sample Number[b]	As	Cd	Cr	Cu	Fe	Mg
A/622	< 0.003	< 0.03	< 0.002	< 0.2	< 0.08	< 4
B/623	< 0.001	< 0.02	< 0.002	< 0.4	< 0.07	< 4
C/726	< 0.017	< 0.03	< 0.003	< 0.3	< 0.09	< 4
A/727	< 0.002	< 0.02	< 0.002	< 0.4	< 0.06	< 5
D/735	< 0.0006	< 0.007	< 0.0008	< 0.3	< 0.05	< 4
C/746	0.15	< 0.1	0.96	< 7	20	< 80
E/747	0.17	< 0.09	0.80	< 7	18	< 100
F/748	0.33	< 0.1	0.92	< 7	35	< 70
F/749	0.15	< 0.09	0.80	< 4	15	< 60
G/750	0.11	< 0.07	0.077	< 8	13	< 100
H/790	0.20	< 0.06	0.037	< 8	12	< 50
H/791	0.21	< 0.05	0.035	< 2	14	< 30
H/792	0.13	< 0.06	0.030	< 2	8.8	< 50
I/833	0.001	< 0.02	< 0.003	< 0.5	< 0.05	< 5
J/938	0.0037	< 0.03	< 0.003	< 0.2	< 0.07	< 4
K/972	< 0.002	< 0.02	< 0.002	< 0.5	< 0.07	< 5
B/979	< 0.001	< 0.02	< 0.003	< 0.2	< 0.06	< 5

[a] In $\mu g/mL$.

"Typical" emission factors for metals cannot be derived from baseline characterization of auto exhaust by dynamometer tests, as performed by EPA, since attempts are made to keep variability of additives, oils, and lubricants to a minimum. Emphasis is placed rather on the effect of emissions as a function of variations in operating conditions. The data cited in Table X reflect this because test cycles identified as FTP, HWFET, and CFS differ significantly in the average speed (19.9, 49.9, and 35.0 mph, respectively) and in the extent of variability in operating mode (acceleration, deceleration, and cruise).

The relative contribution of mobile source emissions to atmospheric lead levels over the next decade is expected to decrease slightly for two reasons: (1) the Federal government's imposition of a phasedown of lead content in gasoline, as called for in regulations promulgated in September 1976 (43) and (2) the incompatibility of leaded gasoline and the catalytic converter introduced in late 1974 for control of regulated (HC, CO, NO_x) pollutants (44, 45).

Emissions of lead from mobile sources differ significantly from stationary source emissions in that automotive lead is closely associated with bromine and chlorine co-added as scavengers. In general, two principal types of lead compounds have been identified in vehicular exhaust: the normal lead halide PbClBr and binary complex compounds

Fuel Oils, Elemental Analysis[a]

Mn	Ni	Sb	Se	V	Zn
< 0.005	< 0.02	< 0.0002	0.0084	< 0.001	0.089
< 0.004	< 0.01	< 0.0002	< 0.003	< 0.003	0.044
< 0.005	< 0.01	< 0.0002	0.012	< 0.003	0.11
< 0.004	< 0.01	< 0.0001	0.023	< 0.003	0.075
0.004	< 0.01	< 0.0001	< 0.0009	< 0.002	0.56
0.24	11	0.0031	0.11	24.2	3.4
0.14 ± 0.04	16	0.0074	0.14	21.3	1.7
0.20	14	0.0070	0.15	10.2	3.0
0.18	12	0.0035	0.15	9.5	1.8
< 0.07	13	0.0015	0.11	38.1	1.0
0.05 ± 0.02	4.6	0.0016	0.034	5.9	0.31
0.08 ± 0.04	6.0	0.0016	0.092	6.6	0.83
< 0.05	4.0	0.0011	0.058	5.9	0.21
< 0.003	< 0.02	< 0.0001	< 0.004	< 0.003	0.11
< 0.004	< 0.02	< 0.0001	0.045	0.004 ± 0.002	0.23
< 0.004	< 0.01	< 0.0001	0.0037	< 0.003	0.28
< 0.004	< 0.04	< 0.002	< 0.003	< 0.003	4.1

[b] Brand identification given in Ref. *64*, p. 14.

of this halide and ammonium chloride, as well as minor amounts of other compounds. Studies of vehicles operated under varying conditions show that emission of PbClBr is favored at higher temperatures while the emission of ammonium halide compounds is favored under lower temperature conditions (*46*). Particle size appears to be temperature dependent as well as composition dependent, but over the lifetime of the vehicles about 35% of lead exhausted is as fine (< 5 μm in diameter) particulate and about 55% as coarse (*47*). It is important to note that emission factors such as those listed in Table X are derived from dynamometer tests in which exhaust is characterized from dilution tunnel samples and do not reflect total emissions because only that particulate which would normally remain suspended in the atmosphere ($\sim$ < 10 μ) is characterized and quantified. Variability in rate of lead particulate (total and/or airborne) exhausted, composition, and size distribution is substantial and depends on a number of operating condition variables.

Manganese, which along with lead is emitted to the atmosphere mainly through combustion of fuel rather than by wear or ablation, is being promoted as an alternative anti-knock in lead-free gasoline. The manganese additive, methylcyclopentadienylmanganesetricarbonyl (MMT) is, on a weight-of-metal basis, more effective than lead; its recommended concentration is 0.125 g/gal (*48*). This is, of course, much

Table XII. Selected Multigrade (10W-40)

Brand Code[b]	As	Cd	Cr	Cu	Fe	Mg
A/255	0.086	< 0.09	< 0.2	< 1	< 3	460
B/258	0.26	< 0.09	< 0.1	< 1	< 3	< 20
C/374	0.24	< 0.2	< 0.09	< 1	< 3	< 20
D/375	0.043	< 0.09	< 0.2	< 1	7.9 ± 2.0	1160
A/572	0.096	< 0.09	< 0.09	< 1	< 3	410
B/573	0.16	< 0.2	< 0.09	< 1	< 3	< 20
D/574	< 0.004	< 0.09	0.051 ± 0.036	< 1	< 5	16 ± 10
D/574	0.05	< 0.02	0.28	< 0.2	4.8	20 ± 7
B/620	0.29	< 0.09	< 0.08	< 1	2.4 ± 1.7	< 20
D/660	0.017 ± 0.004	< 0.2	< 0.08	< 1	5.2 ± 1.3	50 ± 10
B/655	< 0.02	< 0.2	< 0.07	< 3	< 5	< 60
B/707	0.21	< 0.09	< 0.09	< 1	3.8 ± 1.5	< 20
F/712	0.034 ± 0.007	< 0.2	< 0.09	< 3	4.6 ± 1.4	< 60
F/712	0.025	< 0.02	< 0.07	< 0.2	4.1	8 ± 4
D/752	< 0.02	< 0.2	< 0.06	< 1	< 5	40 ± 10
A/767	0.009 ± 0.003	< 0.009	< 0.009	< 1	2.4 ± 1.2	670
A/767	0.017	0.42	< 0.08	< 0.2	< 0.5	770
D/848	0.30 ± 0.005	< 0.2	< 0.09	< 1	< 3	1170
B/850	0.22	< 0.2	< 0.06	< 1	< 2	< 20
E/864	0.035 ± 0.010	< 0.2	< 0.09	< 1	7.3 ± 1.5	< 20
B/870	0.16	< 0.2	0.18 ± 0.05	< 1	3.5 ± 1.4	< 20
F/920	< 0.2	< 0.2	< 0.09	< 1	4.6 ± 1.4	< 20
A/922	0.04	< 0.2	< 0.06	< 1	3.1 ± 1.6	390

[a] In $\mu g/mL$.

lower than a typical lead additive level such as the 2.8 g/gal used in test runs to provide the lead emission factors shown in Table X. Vehicles tested with gasoline containing the 0.125 mg/gal level of MMT have produced particulate manganese emissions ranging from 1.4 to 16 mg/mile (49). This particulate is reported by Ethyl Corporation to be in the form of Mn_3O_4 (48). However, evidence is accumulating that other manganese species can be present. A study by EPA reported very low but detectable levels of the parent compound (MMT) in exhaust, 0.005 mg/mile at most (50). The size distribution of manganese appears to be similar to that of lead exhaust particulate and, like lead particulates, is also temperature dependent (49).

Other environmentally important metals commonly found in automotive exhaust, such as iron, copper, and nickel, result most probably from wear or ablation rather than from combustion since very small amounts of these metals have been identified in fuels, fuel additives, or motor oils (40). Aluminum, for instance, though not an environmentally important metal, is an exhaust component commonly identified in exhaust from

Motor Oils, Elemental Analysis[a]

Mn	Ni	Sb	Se	V	Zn
0.18	< 4	0.26	< 0.08	< 0.01	1360
0.23	< 4	0.16	< 0.09	< 0.09	1310
0.25	< 3	0.021 ± 0.003	< 0.08	< 0.01	1700
0.10 ± 0.03	< 4	< 0.007	< 0.08	< 0.01	1880
0.14 ± 0.02	< 3	0.25	< 0.07	< 0.01	1450
0.22	< 3	0.084	< 0.08	< 0.01	1290
0.21	< 3	0.009 ± 0.003	< 0.06	< 0.01	1040
0.19	< 1	0.016	< 0.03	< 0.01	1700
0.30	< 3	0.022 ± 0.003	< 0.06	< 0.01	1240
0.22	< 4	0.009 ± 0.003	< 0.06	< 0.01	1500
0.08 ± 0.03	< 2	0.005 ± 0.002	< 0.06	0.04 ± 0.02	1120
0.16	< 3	0.064	< 0.06	0.04 ± 0.02	1310
0.08 ± 0.02	< 3	0.005 ± 0.002	< 0.06	< 0.03	1090
0.10	< 1	0.005	< 0.02	< 0.01	1100
0.19	< 3	0.13 ± 0.005	< 0.06	< 0.01	1200
0.10 ± 0.02	< 3	0.008 ± 0.003	< 0.06	< 0.01	1100
< 0.04	< 1	0.007	< 0.02	0.006 ± 0.002	1300
0.21	< 3	0.006 ± 0.002	< 0.09	< 0.01	1990
0.16	< 3	0.43	< 0.06	0.016 ± 0.004	1230
0.08 ± 0.02	< 3	< 0.005	< 0.06	< 0.01	1240
0.30	< 2	0.066	< 0.04	< 0.01	1280
0.04 ± 0.02	< 3	0.008 ± 0.003	< 0.06	< 0.01	1180
0.17	< 2	0.017 ± 0.003	< 0.04	< 0.01	1140

[b] Brand identification given in Ref. *64*, p. 7.

vehicles with aluminum blocks, especially those with low mileage (*39*). Nickel emissions, as shown in Table X, have been most prominent in exhaust from vehicles equipped with nickel-containing reduction catalysts (*see* dual catalyst vehicle data). An emission rate of 7.05 mg/mile Ni was found in dynamometer tests of a low-mileage vehicle so equipped, which was equivalent to 40% of the total emitted particulate (*41*). The rate decreased rapidly, however, with mileage accumulation.

Other environmentally important metals, though not detected to date in automotive exhaust, have been found in gasoline, in additives to gasoline, and in motor oils (*see* Tables XI, XII, and XIII). In addition to the metals listed in the tables, the presence of cadmium in gasoline and motor oil has been reported (*51*), and it has been suggested that mercury may enter gasoline in the process of distillation (*52*). If relative concentrations in gasoline should increase or if significant improvements are made in the sensitivity of analytical techniques for exhaust particulate characterization, metals from these sources may be of concern in the future. If these metals, however low their concentrations, could be shown to be

Table XIII.   Selected Fuel Additives as Marketed

Additive Type	Brand Code[b]	As	Cd	Cr	Cu	Fe	Mg
gas	006	< 0.001	< 0.03	< 0.009	< 1	< 0.9	< 20
gas	006	< 0.001	< 0.02	0.006	—	< 0.04	—
gas	013	< 0.001	< 0.1	0.15 ± 0.02	< 1	535	< 20
gas	013	< 0.007	< 0.9	0.10	—	560	—
gas	016	0.18	< 0.06	< 0.02	< 1	< 0.8	< 20
gas	016	0.23	< 0.04	< 0.01	—	< 0.2	—
gas	477	< 0.004	< 0.1	< 0.04	< 1	< 0.9	< 20
oil	478	< 0.005	< 0.06	< 0.02	< 3	< 0.8	< 60
oil	479	0.018 ± 0.011	< 0.2	< 0.09	< 1	< 0.3	< 20
gas	480	< 0.02	< 0.2	< 0.009	< 40	< 0.9	< 500
gas	481	< 0.004	< 0.03	< 0.004	< 1	< 0.8	< 20
gas	482	< 0.02	< 0.06	< 0.009	1.3 ± 0.5	< 0.9	< 20
gas	483	< 0.002	< 0.04	< 0.02	< 1	< 0.4	< 20
oil	484	< 0.06	< 0.5	< 0.3	< 6	31.6	4740
gas	485	< 0.008	< 0.05	< 0.02	< 1	6.6	< 20
gas	486	< 0.003	< 0.06	< 0.008	< 1	< 0.8	< 20
gas	487	< 0.005	< 0.06	< 0.009	< 1	< 0.9	< 20
oil	488	< 0.05	< 0.3	< 0.3	< 1	15.3	< 20
oil	489	< 0.003	< 0.06	< 0.009	< 1	< 0.8	< 20
gas	490	< 0.001	< 0.05	< 0.02	< 1	85	< 20
oil	491	< 0.001	< 0.05	< 0.02	< 1	2 ± 0.14	< 20
gas	492	< 0.005	< 0.06	< 0.02	< 1	1 ± 0.6	< 20

[a] In $\mu g/mL$.

both widespread and consistent, the contribution from mobile sources to atmospheric levels of these metals would be relatively significant because of the ubiquity of the source.

In addition to ablation products such as nickel, copper, and iron now commonly present in exhaust, there is the potential occurrence in exhaust particulate of metals in use or considered for use as catalyst materials. Their detection would depend upon both increased use and/or increased analytical sensitivity. Efforts to quantify by routine methods emissions of platinum and palladium, now in widespread use in the oxidation catalyst, have failed to substantiate the presence of these metals in exhaust. An emission factor of 3.1×10^{-5} g/mile for platinum has been estimated (53). Other metals which have been considered for use in automotive catalysts include ruthenium and vanadium.

Tire wear is an ablation product of vehicle operation not found in exhaust which can be considered significant. Tire wear emissions have been estimated at 7 million ton/year in the United States (54). Both zinc and cadmium are components of tire rubber, and tire debris is a significant source of atmospheric emissions of these metals. Zinc comprises approximately 1.5% of tire dust by weight. In Los Angeles, tire dust is one of two major sources of atmospheric zinc, the other being metallurgical emissions (55). Cadmium emissions from tire debris was estimated at

for Consumer Purchase and Use, Elemental Analysis[a]

Mn	Ni	Sb	Se	V	Zn
< 0.01	< 0.09	< 0.0009	< 0.0009	< 0.01	0.14 ± 0.02
< 0.003	< 0.02	< 0.0001	< 0.002	< 0.003	0.35
2.60	< 0.7	< 0.003	< 0.03	0.036 ± 0.010	4.1
2.71	< 0.1	0.002	< 0.003	0.016 ± 0.004	4.2
< 0.01	< 0.2	0.006 ± 0.002	< 0.02	< 0.01	11
0.012 ± 0.002	< 0.2	0.003	< 0.007	< 0.01	15
0.016 ± 0.004	< 0.3	< 0.0009	< 0.04	< 0.01	2.5
< 0.03	< 0.2	< 0.0009	< 0.04	< 0.03	8.8
< 0.01	< 4	< 0.009	< 0.09	< 0.01	3.70
< 0.5	< 0.09	< 0.0009	< 0.09	< 2	0.24
< 0.1	< 0.08	< 0.0008	< 0.004	< 0.01	0.17
0.052 ± 0.008	< 0.3	< 0.0009	< 0.004	< 0.01	2.4
< 0.01	< 0.2	< 0.002	< 0.02	< 0.01	0.19
< 0.01	< 9	0.06 ± 0.02	< 0.2	< 0.05	6390
0.039 ± 0.006	< 0.2	< 0.002	< 0.02	< 0.01	3.8
< 0.01	< 0.2	< 0.002	< 0.004	< 0.01	2.6
< 0.01	< 0.2	< 0.002	< 0.004	< 0.01	4.9
< 0.086	< 5	0.035 ± 0.009	< 0.2	< 0.01	4050
< 0.01	< 0.2	< 0.002	< 0.004	0.19	0.34
0.19	< 0.2	0.013	< 0.02	< 0.01	0.85
< 0.01	< 0.4	< 0.002	< 0.02	< 0.01	23
< 0.01	< 0.4	< 0.099	0.60	< 0.01	13

[b] Brand identification given in Ref. *64*, p. 4.

11,000 lb in the United States in 1968. Though the amount is very small, tire debris ranked seventh in a listing of 16 major sources of atmospheric emissions of cadmium in the United States (*51*).

With respect to contributions of other mobile sources to the atmospheric burden of trace metals, very few data are available. The National Aeronautics and Space Administration in 1974 published calculated annual average ambient concentrations at or near airports of 49 trace elements attributable to aircraft; most values were "less than" numbers. Of interest, however, are the estimates for the environmentally important metals, titanium, vanadium, and cadmium; these were given as 24 ng/m^3, 0.12 ng/m^3, and < 14 ng/m^3, respectively (*56*). Efforts to characterize aircraft emissions and to develop emissions factors by means of dynamometer tests are under way. One such experiment is being performed at Pratt and Whitney under contract to the Environmental Protection Agency; results from this study should be available in mid-1978 (*57*).

Relating Airborne Metals to Sources

It is possible to distinguish whether airborne metals arise from crustal sources, e.g., windborne soil and rocks, or from anthropogenic sources by comparing the elemental concentration patterns of airborne particles to that of crustal material (6, 58). Enrichment factors can be calculated by determining the ratio of the given metal in air and crustal material to that of a major nonvolatile element such as silicon, aluminum, or iron:

$$EF = \frac{(C_x/C_{Al})_{\text{Air}}}{(C_x/C_{Al})_{\text{Crust}}}$$

where: EF = enrichment factor; C_x = concentration of metal in air or crustal material; and C_{Al} = concentration of Al or Si or Fe in air or crustal material.

An enrichment factor of one or somewhat greater indicates that the airborne metal probably has a crustal origin while a large enrichment factor probably denotes an anthropogenic origin. Gordon and Zoller (58) compiled enrichment factors of airborne particles in Boston, Indiana, and San Francisco compared with crustal abundances. Low or modest enrichments (EF = 1–10) were found for magnesium, chromium, manganese, and iron compared with aluminum. An enrichment factor of more than 100 was calculated for vanadium in Boston, attributed to the combustion of residual oils; low vanadium enrichment factors were found for San Francisco and Indiana. Metals that exhibited high enrichment factors (greater than 100) included nickel, copper, zinc, arsenic, cadmium, tin, antimony, lead, and selenium, which the authors attributed to coal and oil combustion and incineration.

Other workers carried this technique one step farther with the use of sensitive, multi-element analytical methods. King et al. (59) applied instrumental neutron activation, emission spectroscopy, and combustion techniques to measure the concentration of 60 chemical elements at 16 air monitoring sites in Cleveland. Efforts were made to compare the airborne trace element profile with the trace element profiles of various sources, assuming that each type of source emits a characteristic distribution of a set of elements. Using a combination of pair-wise correlation statistics, cluster analysis algorithms, enrichment factors, and pollutant concentration sources, Neustadter et al. (60) were able to identify many of the specific sources of environmentally important metals in Cleveland: earth's crust (titanium, vanadium, and tin), automotive exhaust (lead), metallurgy (iron and chromium), and specific industries (antimony and mercury). Friedlander (61) found that greater than 72% of the aerosol in Pasadena air originated from sea salt, soil, auto exhaust, fuel oil, fly ash, and cement dust. Using similar approaches for Chicago, Winchester (9) attributed arsenic, chromium, tin, titanium, nickel, and vanadium to coal, coke, and oil combustion; iron, manganese, copper, and zinc from iron and steel production; and lead and cadmium from gasoline and coal. John et al. (62) classified San Francisco aerosol as marine origin (sodium, chlorine, and bromine), soil origin (iron, manganese, and chromium) which accounted for most of the aerosol, and anthropogenic origin (antimony, selenium, zinc, and mercury).

Hammerle and Pierson (63) took this approach one step farther by measuring the particle-size distribution as well as the concentration of trace elements by x-ray fluorescence analysis (including the environmentally important metals) in Pasadena. They classified nickel and zinc as small particle elements and titanium, manganese, and iron as large particle elements, with vanadium falling in an intermediate size range. These results are in general agreement with the findings of others (6, 7, 8) which indicate that lead, vanadium, and zinc are associated with particles submicrometer in diameter; iron and possibly magnesium with large particles greater than 2 μm in diameter; and copper, nickel, chromium, and tin associated with intermediate size ranges. Hammerle and Pierson (63) applied an inter-element correlation and functional model to these data and concluded that soil is the main source of titanium, manganese, and iron, and that gasoline is the main source of lead while vanadium, nickel, and zinc sources were not identified.

It is apparent that these "fingerprint" techniques relating particle sizes as well as concentration data to source components need further refining. However, it should be possible eventually to identify the origin of airborne particles provided sufficient chemical data become available. Recently, lead isotopic ratios have been used to estimate the contribution of blood lead levels originating from gasoline combustion. Perhaps this approach can be further refined to better characterize the sources of other airborne metals.

Summary and Conclusions

The environmental scientist has at his disposal a variety of sensitive, multi-elemental analytical methods that can lead to a massive amount of data on airborne metals. Optimum use of these tools for environmental monitoring calls for focusing resources only on those metals that are environmentally important. Considerations of toxicity along with their ability to interact in the air, leading to the formation of secondary pollutants, and their presence in air have led to the identification of 17 environmentally important metals: nickel, beryllium, cadmium, tin, antimony, lead, vanadium, mercury, selenium, arsenic, copper, iron, magnesium, manganese, titanium, chromium, and zinc. In addition to the airborne concentration, the particle size of environmentally important metals is perhaps the major consideration in assessing their importance.

Ocean spray, volcanic activity, and windblown soil and dust are the major natural sources of environmentally important metals, including magnesium, manganese, and iron. However, most of these emissions are associated with large airborne particles which lessen their direct importance to human health. The major stationary emission sources are smelter/metallurgical processing, coal and oil combustion, municipal incineration, and cement manufacturing. Of these, smelter/metallurgical processing account for 43% of the total airborne Cd, 68% Cr, 84% Cu, 57% Mn, 11% Ni, 3% Be, and 2% Pb. Coal combustion accounts for 88% of the total airborne Be, 9% Cr, 7% Cu, 11% Mn, and 9% Ti; oil combustion: 6% Be, 83% Ni, and 90% V; incineration accounts for 52% of the total airborne Cd and 1% of the Pb. Combustion processes are especially important because of the preferential concentration of volatile metals (lead, antimony, cadmium, selenium, arsenic, zinc, nickel, and chromium) in fine particles; mercury appears to be almost totally volatilized into the gaseous phase. Mobile sources, specifically automotive exhaust, contribute the major quantity of airborne lead and lesser amounts of manganese, iron, copper, nickel, and zinc. Tire wear accounts for 7 million tons of particulate each year in the United States, resulting in the emissions of significant amounts of zinc and cadmium.

Efforts have been made over the past several years to identify airborne metals with specific emission sources by the use of enrichment factors, and by comparing the trace element profile of airborne particles with characteristic components in particulate matter from various sources. Measurements of particle size and the application of sophisticated statistical techniques should increase the accuracy of these "fingerprinting" approaches.

Literature Cited

1. "Threshold Limit Values for Chemical Substances in Workroom Air Adopted by ACGIH for 1977," American Conference of Governmental Industrial Hygienist, Cincinnati, OH, 1977.
2. Schroeder, H. A., "Metals in the Air," *Environment* (1971) **13**, 20.
3. Wilcox, S. L., "Presumed Safe Ambient Air Quality Levels for Selected Potentially Hazardous Pollutants," prepared by the Mitre Corp., Washington, D.C., under EPA Contract No. **68-01-0438** (1973).
4. "Study of the Nature of the Chemical Composition of Particles Collected from Ambient Air," prepared by Battelle Memorial Institute, Columbus, OH, under EPA Contract No. **CPA 22-69-153** (1970).
5. Lee, R. E., Jr., Goranson, S. S., Enrione, R. E., Morgan, G. B., "National Air Surveillance Cascade Impactor Network. II. Size Distribution Measurements of Trace Metal Components," *Environ. Sci. Technol.* (1972) **6**, 1025.
6. Lee, R. E., Jr., von Lehmden, D. J., "Trace Metal Pollution in the Environment," *J. Air Pollut. Control Assoc.* (1973) **23**, 853.
7. Gladney, E. S., Zoller, W. H., Jones, A. G., Gordon, G. E., "Composition and Size Distributions of Atmospheric Particulate Matter in Boston Area," *Environ. Sci. Technol.* (1974) **8**, 551.

8. Cunningham, P. T., Johnson, S. A., Yang, R. T., "Variations in Chemistry of Airborne Particulate Material with Particle Size and Time," *Environ. Sci. Technol.* (1974) **8,** 31.
9. *National Emissions Report* EPA-450/2-76-007, U.S. Environmental Protection Agency, Research Triangle Park, NC (1976).
10. Winchester, J. W., "Trace Metal Associations in Urban Airborne Particulates," *Bulletin of the American Meteorological Society* (1973) **51,** 94.
11. Shannon, L. J., Gorman, P. G., Reichel, M., "Particulate Pollutant System Study Vol. II: Fine Particle Emissions," prepared by Midwest Research Institute, Kansas City, KS, under EPA Contract No. **CPA-22-69-104** (1971).
12. "Investigations of Fugitive Dust: Sources, Emissions, and Controls," prepared by PEDCo Environmental, Cincinnati, OH, under EPA Contract No. **68-02-0044** (1973).
13. "Inadvertent Climate Modification: Report of the Study of Man's Impact on Climate (SMIC)," p. 189, MIT Press, Cambridge, MA, 1971.
14. Marchesani, V. J., Towers, T., Wohlers, H. C., "Minor Sources of Air Pollutant Emissions," *J. Air Pollut. Control Assoc.* (1970) **20,** 19.
15. Willeke, K., Whitby, K. T., "Atmospheric Aerosols: Size Distribution Interpretation," *J. Air Pollut. Control Assoc.* (1975) **25,** 529.
16. Hobbs, P. V., Radke, L. F., Stith, J. L., "Eruptions of the St. Augustine Volcano: Airborne Measurements and Observations," *Science* (1977) **195,** 87.
17. Mason, B., "Principles of Geochemistry," Wiley, New York, 1966.
18. Mroz, E. J., Zoller, W. H., *Science* (1975) **190,** 461.
19. Vinogradov, A. P., "The Geochemistry of Rare and Dispersed Chemical Elements in Soil," Consultants Bureau Inc., New York, 1959.
20. Andrew-Jones, D. A., *Miner. Ind. Bull.* (1968) **11,** 1.
21. Chester, R., Stoner, J. H., "Marine Chemistry" (1974) **2,** 157.
22. Beauford, W., Barber, J., Barringer, A. J., "Release of Particles Containing Metals from Vegetation into the Atmosphere," *Science* (1977) **196,** 571.
23. Flinn, J. E., Reimers, R. S., "Development of Predictions of Future Pollution Problems," EPA Report No. **600/5-74-005** (1974) pp. 36–38.
24. Faoro, R. B., McMullen, T. B., "National Trends in Trace Metals in Ambient Air," EPA Report No. **450/1-77-003** (1977).
25. Rubel, F. H., "Incineration of Solid Wastes," pp. 56–57, Noyes Data Corp., Park Ridge, NJ, 1974.
26. Kirsch, H., "Composition of Dust in the Waste Gases of Incineration Plants," presented at the First International Conference and Technical Exhibition (IEE Catalog No. **75 CH1008-2 CRE**), Montreux-Switzerland (1975).
27. Lee, R. E., Jr., Crist, H. L., Riley, A. E., MacLeod, K. E., "Concentration and Size of Trace Metal Emissions from a Power Plant, a Steel Plant, and a Cotton Gin," *Environ. Sci. Technol.* (1975) **9,** 643.
28. Klein, D. H., Andren, A. W., Carter, J. A., Emery, J. F., Feldman, C., Fulkerson, W., Lyon, W. L., Ogle, J. C., Talmi, Y., Van Hook, R. I., Bolton, N., "Pathways of Thirty-Seven Trace Elements Through Coal-Fired Power Plants," *Environ. Sci. Technol.* (1975) **9,** 973.
29. Toca, F. M., Cheever, C. L., Berry, C. M., "Lead and Cadmium Distribution in the Particulate Effluent from a Coal-Fired Boiler," *J. Am. Ind. Hyg. Assoc.* (1973) **34,** 396.
30. "Coal-Fired Power Plant Trace Element Study," Report prepared for the U.S. Environmental Protection Agency by the Radian Corporation, Austin, TX (1975).
31. Francher, J. R., "Trace Element Emissions from the Combustion of Fossil Fuels," *Proc. Conf. Cycling and Control of Metals*, sponsored by EPA, NSF, and Battelle Memorial Laboratories, Cincinnati, OH, 1972.

32. Ruch, R. R., Gluskota, H. J., Shimp, M. F., "Occurrence and Distribution of Potentially Volatile Trace Elements in Coal: A Final Report," Environ. Geol. Notes, Ill. State Geol. Surv. (1974) **72**.

33. Bennett, R. L., Knapp, K. T., "Chemical Characterization of Particulate Emissions from Oil-Fired Power Plants," *National Conference on Energy and the Environment, 4th, Cincinnati, OH, 1976.*

34. Knapp, K. T., Conner, W. D., Bennett, R. L., "Physical Characterization of Particulate Emissions from Oil-Fired Power Plants," *National Conference on Energy and the Environment, 4th, Cincinnati, OH, 1976.*

35. Natusch, D. F. S., Wallace, J. R., Evans, C. A., Jr., "Toxic Trace Elements: Preferential Concentration in Respirable Particles," *Science* (1974) **183**, 202.

36. "Data File on Nationwide Emissions," U.S. Environmental Protection Agency, Monitoring and Data Analysis Division, Office of Air Quality Planning and Standards, Research Triangle Park, NC, 1976.

37. Bradow, R. L., ESRL, EPA, Research Triangle Park, NC, 1977.

38. "Effect of Fuel Additives on the Chemical and Physical Characteristics of Particulate Emissions in Automotive Exhaust," EPA Contract No. **CPA-22-69-145**, Dow Chemical Co. (1970).

39. Gibbs, R., "Sulfate and Particulate Emissions from In-Use Catalyst Vehicles," Interim Progress Report, EPA Research Grant No. **R803520**, New York State Dept. of Environmental Conservation, Albany, NY (1976).

40. Gabele, P., Braddock, J., Bradow, R., "Characterization of Exhaust Emissions from Lean Burn, Rotary, and Stratified Charge Engines," SAE Paper No. **770301** (1977).

41. Gabele, P. A., Braddock, J. N., Black, F. M., Stump, F. D., Zweidinger, R. B., "Characterization of Exhaust Emissions from a Dual Catalyst-Equipped Vehicle," U.S. Environmental Protection Agency, Research Triangle Park, NC, EPA Report No. **EPA-600-2-77-068** (1977).

42. Braddock, J., Bradow, R., "Emissions Patterns of Diesel-Powered Passenger Cars, Part I," SAE Paper No. **750682** (1975).

43. "Regulation of Fuels and Fuel Additives, Control of Lead Additives in Gasolines," *Fed. Regist.* (1976) **41**, 42675.

44. Moran, J. B., "Lead in Gasoline, Impact of Removal on Current and Future Automotive Emissions," presented at Air Pollution Control Association, Denver, CO (1974).

45. Weaver, E. E., "Effects of Tetraethyl Lead on Catalyst Life and Efficiency in Customer Type Operation," presented at International Automotive Engineering Congress, SAE, Detroit, MI, SAE Paper No. **690016** (1969).

46. Hirschler, D. A., Gilbert, L. F., Lamb, F. W., Niebylski, L. M., "Particulate Lead Compounds in Automobile Exhaust Gas," *Ind. Eng. Chem.* (1957) **49**, 1131.

47. ter Haar, G. L., Lenane, D. L., Hu, J. H., Brandt, M., "Composition, Size, and Control of Automotive Exhaust Particulates," *J. Air Pollut. Control Assoc.* (1972) **22**(39).

48. "Methylcyclopentadienylmanganesetricarbonyl (MMT), An Antiknock Agent for Unleaded Gasoline, Status Report," ER-452, Ethyl Corp., Ferndale, MI (1974).

49. Moran, J. B., "The Environmental Implications of Manganese as an Alternate Antiknock," presented at Automotive Engineering and Manufacturing Meeting, SAE, Detroit, MI, SAE Paper No. **750926** (1975).

50. "Effect of Gasoline Additives on Gaseous Emissions," U.S. Environmental Protection Agency, Research Triangle Park, NC, Report No. **EPA-650/2-75-014** (1974).

51. Fleischer, M., Sarofim, A. F., Fassett, D. W., Hammond, P., Shacklette, H. T., Nisbet, I. C. T., Epstein, S., "Environmental Impact of Cadmium, a Review by the Panel on Hazardous Trace Substances," *Environ. Health Persp.* (1974) **7**, 252–323.
52. Yamaguchi, M., Shimojo, N., "Methyl Mercury Derivatives in Engine Exhausts and Its Synthesis," *J. Jpn. Soc. Air Pollut.* (1975) **10**, 413.
53. *Proceedings, Catalyst Research Program Platinum Research Review Conference*, Rougemont, NC, sponsored by the Health Effects Research Laboratory, U.S. Environmental Protection Agency, Research Triangle Park, NC (1975).
54. "A Survey of Automotive Emissions," Naval Research Laboratory, Washington, D.C., Report No. **NRL-MR-2346** (1971).
55. Huntzicker, J. J., Davidson, C. I., Friedlander, S. K., Keck, W. M., "The Flow of Trace Elements Through the Los Angeles Basin: Zn, Cd, and Ni," in "Abstracts of Papers," 167th National Meeting, ACS, March–April 1974, ENVT 48.
56. Fordyce, J. S., Sheihley, D. W., "Estimate of Contribution of Jet Aircraft Operations to Trace Element Concentration at or Near Airports," National Aeronautics and Space Administration, Washington, D.C., Report No. **NASA TM X-3054** (1974).
57. Bradow, R. L., Environmental Sciences Research Laboratory, U.S. Environmental Protection Agency, Research Triangle Park, NC, personal communication.
58. Gordon, G. E., "Study of the Emissions from Major Air Pollution Sources and Their Atmospheric Interactions," University of Maryland, College Park, prepared under NSF Grant No. **GI-36338X**, pp. 17-21 (1974).
59. King, R. B., Fordyce, J. J., Antoine, A. C., Liebecki, H. F., Neustadter, H. E., Sidik, S. M., "Elemental Composition of Airborne Particulates and Source Identification: An Extensive One-Year Survey," *J. Air Pollut. Control Assoc.* (1976) **26**, 1079.
60. Neustadter, H. E., Fordyce, J. J., King, R. B., "Elemental Composition of Airborne Particulates and Source Identification: Data Analysis Techniques," *J. Air Pollut. Control Assoc.* (1976) **26**, 1079.
61. Friedlander, S. K., "Chemical Element Balances and Identification of Air Pollution Sources," *Environ. Sci. Technol.* (1973) **7**, 235.
62. John, W., Kaifer, R., Rahn, K., Wesolowski, J. J., "Trace Element Concentrations in Aerosols from the San Francisco Bay Area," *Atmos. Environ.* (1973) **7**, 107.
63. Hammerle, R. H., Pierson, W. R., "Sources and Elemental Composition of Aerosol in Pasadena, California, by Energy-Dispersive X-Ray Fluorescence," *Environ. Sci. Technol.* (1975) **9**, 1058.
64. "Annual Catalyst Research Program Report, Appendices, Vol. II," EPA Report No. **600/3-75-010c** (1975).

RECEIVED November 21, 1977. This report has been reviewed by the Health Effects Research Laboratory, U.S. Environmental Protection Agency, and approved for publication. Mention of trade names or commercial products does not constitute endorsement or recommendation for use.

12

Trace Metal Uptake by *Bacillus subtilis* Strain 168

J. J. DULKA and T. H. RISBY[1]

Department of Chemistry, Pennsylvania State University,
University Park, PA 16802

P. E. KOLENBRANDER

Department of Microbiology and Cell Biology, Pennsylvania State University,
University Park, PA 16802

The uptake of aluminum, cadmium, chromium, cobalt, copper, iron, lead, magnesium, manganese, molybdenum, nickel, silver, tin, and zinc by B. subtilis *Strain 168 is reported. These data were obtained during the lag phase, exponential phase, stationary phase, and the sporulation phase of the maturation cycle of this bacterial strain. Non-flame atomic absorption spectrometry was the method of analysis for all the metals except calcium, which was determined by flame atomic absorption spectrometry. The complete microbiological and analytical procedures are described. Uptake curves as a function of moles per cell, of moles per dry weight of a cell, and of percent available are reported. The data show that these metals seem to be required for growth. No attempts were made to postulate the roles played by these metals.*

The importance of metals in biological systems has been apparent for a number of years, and it has been established that certain metals are required for biological functions, others are toxic, and the remainder have no known requirements (*1*). Metals have roles in enzyme activities, nerve signal transport, proteins, hemes, and cellular structure (skele-

[1] Author to whom correspondence should be sent. Current address: Department of Environmental Health Services, Johns Hopkins School of Hygiene and Public Health, 615 N. Wolf St., Baltimore, MD 21205.

0-8412-0416-0/79/33-172-172$06.25/0

ton). However, even these general statements must be qualified since the oxidation state, chemical identity, and concentration will dictate whether a metal is toxic or required. On the basis of these qualifications it can be stated that the knowledge of the roles that metals play in biological systems (or metabolic processes) are still perfunctory, and yet it is essential that the environmental health impact of metals be written. Current research is aimed at establishing whether metals that are listed as having no known requirement do have biological requirements, and almost daily another enzyme is shown to require a metal for its activity. In the area of toxicity, the data is much less satisfactory since, while it has been established which compounds are toxic to certain laboratory animals (LD_{100} or LD_{50}), it does not follow that since one compound of a particular element is toxic that all compounds of that element are toxic. This idea is shown by consideration of two barium salts; barium chloride is extremely toxic whereas barium sulfate is nontoxic and is used as a diagnostic tool. This difference is related to the solubility of barium chloride as opposed to barium sulfate. In the area of long-term exposure to levels of toxic elements, which as a single dose are nontoxic, the current evidence is sadly lacking. Two possible ways that long-term exposure can be made are the use of tissue cultures and the use of bacteria since these methods will provide the answer much quicker than studies with laboratory animals. The advantages of both methods are that mutagenic properties can be established in a short period of time compared with studies of laboratory animals. This chapter is designed to study the uptake of metals by *B. subtilis* 168 from a complex culture medium. *B. subtilis* is a common soil gram-positive bacterium which grows aerobically and produces endospores. Also, both the genetic and the chemical composition of *B. subtilis* 168 have been well characterized (2, 3, 4).

Most studies involving *Bacillus* have been concerned with determining the macro constituent metals of the cells and spores with most of the attention directed towards establishing whether these metals are involved or required for sporulation. These studies have shown that the following metals are found in either the cells or spores: calcium, potassium, magnesium, sodium, manganese, iron, aluminum, copper, and zinc (5).

Calcium, which is probably the most widely studied element, was thought for a long time to be involved in heat resistance and sporulation via the calcium–dipicolinic acid complex (6). However, more recent studies (7) have shown that it is probably manganese that is responsible for these properties and not the calcium complex. Potassium, iron, magnesium, and manganese are known to be required for normal growth, and several researchers (8–16) have studied these metals during growth and sporulation. However, these metals have never been studied simultaneously during growth. Therefore, the purpose of this study is to determine concomitantly the cellular levels in a single culture during growth. A complex growth medium was used for this study since it supports rapid growth. In addition, the metals contained in the medium were unchanged (i.e., no metals were added or removed from the medium). Particular care was taken to minimize the introduction of metal ions during any stage of the growth, sampling, and analysis of the bacterial samples.

Experimental

Apparatus. A nonflame atomic absorption spectrometer (Varian-Techtron AA-5, Model 63 Carbon Rod Atomizer) with background correction was used for all of the analyses with the exception of calcium. Calcium was determined by flame atomic absorption spectrometry (Varian-Techtron Model 1000).

Procedure. CLEANING PROCEDURES. All of the glassware was soaked in aqua regia (Mallinckrodt, nitric and hydrochloric acids) overnight and then rinsed with doubly distilled de-ionized water. After cleaning, the microbiological glassware was dried and sterilized by placing in an oven at 270°C for 5 hr. The volumetric glassware was stored in doubly distilled de-ionized water until required. The disposable polyethylene pipette tips used to transfer the sample for analysis were cleaned by the method previously described by this laboratory (17).

PREPARATION OF GROWTH MEDIUM. A sterile solution of the chemically undefined growth medium was used to culture the bacteria. This medium, tryptose soy broth with dextrose (TSB, Difco Laboratories), consisted of the following components dissolved in distilled water: pancreatic digest of casein, soy bean peptone, dextrose, sodium chloride, and dipotassium phosphate. This medium was filtered through an ultrafine glass frit and stored for two days to ensure that it was sterile. This procedure was adopted in lieu of auto claving to reduce the possibility of metal contamination.

CULTURING PROCEDURE. To obtain a growth culture that exhibited a negible lag phase, an inoculum culture of mid-to-late exponential phase cells was prepared as follows: a refrigerated stock culture of *B. subtilis* 168 was streaked onto a TSB-agar plate and incubated (8–10 hr at 30°C) and a colony was transferred to the growth medium. This culture was then incubated (37°C) and aerated using a reciprocal shaker (New Brunswick Scientific Co.) until it attained a turbidity of 100 KU red filter (Klett-Summerson Photometric Colorimeter, Klett Manufacturing Co.). At this time, a known aliquot of the culture was transferred to a known volume of fresh growth medium and the resulting culture was incubated (37°C), aerated, and monitored by turbidity measurements.

MICROBIOLOGICAL SAMPLING. *Vegetative Cells.* Cellular samples were obtained at various times after inoculation by sampling a known aliquot of the culture medium. These samples were centrifuged (Sorvall Superspeed RC 2-B, 7000 rpm) at 6°C for 10 min. The supernate was then transferred to the culture medium and the cell pellet was washed by re-suspending it three times in doubly distilled de-ionized water. After each wash the cells were separated by centrifugation and the cells were subsequently lyophilized. The use of distilled water instead of a buffer does represent a potential osmotic shock to the bacterial cells. However, the majority of the cells were viable after these washings, and distilled water does reduce the likelihood of metal contamination.

Spores. The sampling of pure spores presents a problem since they are found in the culture medium with vegetative cells and cellular debris. Therefore, after the cellular material and cells had been separated from the culture medium and washed, the cellular material and cells were re-suspended and separated using a specially designed centrifuge tube. This tube consisted of a capillary tube fused into the bottom of a glass centrifuge tube. The separation was achieved at 1000 rpm followed by 4000 rpm to tightly pack the pellet. Since the spores are more dense than the vegetative cells and cellular debris, the spores and the cellular material were separated in layers. The separated spores were re-suspended and counted with a Petroff–Hausser counting chamber (C. A. Hausser and Son) and were then centrifuged with a normal tube. The resulting pellet, which consisted of approximately 99% spores, was lyophilized.

Cell and Spore Number Determination. The number of cells or spores were determined by the following standard methods at various times after inoculation: Petroff–Hausser counting method with a phase contrast microscopy (Carl Zeiss) and the Coulter counting method (Coulter Counter, Model "B," Coulter Electronics Corp.). No cellular chains were observed.

Cell or Spore Mass Determination. After the cells or spores had been lyophilized they were weighed (Sartorius or Mettler H 20 microbalance). These samples were reweighed after drying for 1 hr at 100°C.

SAMPLE PREPARATION. Two methods of sample preparation were used to obtain a homogeneous solution of the bacterial samples. The first involved wet chemical digestion using nitric acid and hydrogen peroxide. A known weight of the lyophilized sample ($\simeq 40\ \mu g$) was transferred to a Gorsuch digestion apparatus (*18*) and an aliquot of nitric acid (3.0 mL Ultrex, J. T. Baker Co.) was added. The resulting suspension was refluxed for 1 hr, and the mixture was cooled to 40°C. After this time a known volume of hydrogen peroxide (0.5 mL, 30% J. T. Baker Co.) was added dropwise. When all the hydrogen peroxide had been added, the mixture was heated gradually until the excess hydrogen peroxide decomposed, and the solution was then refluxed for 20 min. The resulting clear solution was transferred to a volumetric flask and diluted to volume (10 mL). This whole procedure was repeated without the lyophilized cells or spores, and the resulting solution was used as a blank and to check the carryover.

The second method of sample preparation used a low-temperature asher (Model LTA 600, Tracerlab). A known weight of the lyophilized sample was placed in a quartz boat and ashed with a reduced pressure (0.05 Torr) oxygen electrical discharge until the residue was white (10–24 hr). After this time the residue was transferred with nitric acid (Ultrex) and diluted to volume.

PREPARATION OF STANDARDS. Standard solutions were proposed by diluting certified standards (1000 ppm, Fisher Scientific Co.) with nitric acid (3% Ultrex). These solutions were prepared when required and were not stored.

ATOMIC ABSORPTION SPECTROMETRIC ANALYSIS. The solutions of the cells and spores were analyzed by the method of standard additions. An aliquot (5 L) of the sample was placed in the atomization cell and dried but not ashed nor atomized. Then an aliquot (5 L) of a standard solution was added, and the dry and atomize cycles were performed. The temperature and time for each cycle was found by experiment and is sample dependent. Each analysis was repeated three times and at least three standard additions were made. The resulting peak heights for the atomization step were measured and the concentration of the metal was determined using a weighted least-squares computer program. This procedure was used for the determination of the following elements: aluminum, cadmium, chromium, cobalt, copper, iron, lead, magnesium, manganese, molybdenum, nickel, silver, tin, and zinc.

The determination of calcium by the nonflame methodology was found to be impossible so a nitrous oxide–acetylene flame was used as the atom reservoir.

Results and Discussions

Growth Curve. There are a number of processes occurring during the maturation cycle of a bacterium. A typical growth curve for *B. subtilis* 168 can be divided into several distinct phases: the lag phase, the exponential (logarithmic) phase, the stationary phase, and the sporulation phase. The lag phase is the period of time between inoculation and exponential growth and depends upon the inoculum and the medium. The exponential phase is the period in which the cell divides into daughter cells that in turn divide. This rate is defined as the generation time, which for *B. subtilis* 168 is 35 min under the conditions used in this study. The period of time following exponential growth is termed the stationary phase. This process is believed to occur because of either the exhaustion of an essential nutrient or the accumulation of toxic metabolic products. After the stationary phase, sporogenesis commences (*19*). The sporulation phase has been divided into seven distinct stages (*20*): Stage I involves axial alignment of the nuclear material; Stages II, III, and IV involve the forspore development; and Stages V, VI, and VII involve cortex and spore coat development and endospore maturation. This whole process, once initiated, takes 6–8 hr. Figure 1 shows the relationships between the number of cells in the growth medium and their dry weights as a function of time. Three stages of the growth cycle of *B. subtilis* 168 can be clearly seen from the plot of cell number vs. time, and the dry weight data show that there is considerable weight variation of cell weight during these stages. The lag phase was not observed since the inoculum was obtained during mid-to-late exponential growth. These data were

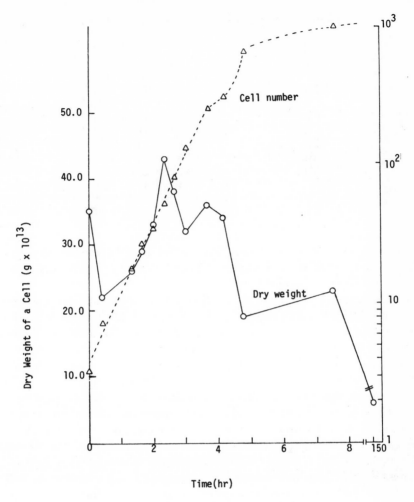

Figure 1. Growth curve

obtained with both the Petroff–Hausser counting method and the Coulter counting method, and the agreement between them was excellent. Therefore, for subsequent measurements, the Coulter counting method was selected since it was more rapid.

Metal Uptake During Growth. The levels of 15 metals (aluminum, cadmium, calcium, chromium, cobalt, copper, iron, lead, magnesium, manganese, molybdenum, nickel, silver, tin, and zinc) were determined in the vegetative cells at various stages during the growth cycle and in the spores. Also these metals were determined in the growth media and the results are shown in Table I. Of the elements investigated, cobalt, molybdenum, and tin were not detected in any of the samples (TSB, cells at

Table I.   Metal Content in Growth Medium

Element	Concentration (mM)
Ag	$(2.7 \pm 0.3) \times 10^{-5}$
Al	$(8.8 \pm 0.8) \times 10^{-2}$
Ca	$(1.63 \pm 0.02) \times 10^{0}$
Cd	$(1.22 \pm 0.06) \times 10^{-5}$
Cr	$(2.42 \pm 0.08) \times 10^{-3}$
Cu	$(1.40 \pm 0.03) \times 10^{-3}$
Fe	$(1.6 \pm 0.1) \times 10^{-2}$
Mg	$(5.5 \pm 0.1) \times 10^{-1}$
Mn	$(2.7 \pm 0.2) \times 10^{-3}$
Ni	$(1.4 \pm 0.1) \times 10^{-3}$
Pb	$(1.8 \pm 0.2) \times 10^{-4}$
Zn	$(5.54 \pm 0.08) \times 10^{-3}$

various growth times or spores), and nickel was detected in the TSB, spores, and one of the cell samples. Therefore, the following discussion will not include these elements.

The results of the analyses for the other 11 elements are presented in three different ways (Figures 2–16). Figures 2–12 show the logarithm of the number of moles of the metal per dry weight of a cell vs. growth time and the relationship between the logarithm of the number of moles of the metal per cell vs. growth time, and Figures 13, 14, 15, 16 show the percentage of the metal found in the cells vs. growth time. From these curves various conclusions can be drawn.

These curves show that aluminum, cadmium, chromium, copper, lead, silver, and zinc were concentrated in the cell during the first 25 min of growth whereas the iron, magnesium, and manganese levels had dropped and the calcium remained constant. These uptakes follow a general trend in which mid-to-late exponential phase inoculum cells change to the early exponential growth cells since the former cells have lower levels of aluminum, cadmium, chromium, copper, lead, silver, and zinc and higher levels of magnesium and manganese than the latter cells. There is no change in the level of calcium in the mid-to-late exponential phase. The data for iron does not follow the same trend but the uptake vs. the release is not as large for iron throughout the growth cycle as for the other elements. These data suggest that when mid-to-late exponential inoculum cells are transferred to fresh growth medium, they begin exponential growth once the metal content of the cells reaches the typical concentrations for the early stage of exponential growth.

Once the cells have reached the exponential phase in the growth cycle, the levels of calcium, magnesium, and manganese increased, iron remained approximately constant, and the other metals decreased (aluminum, cadmium, copper, chromium, lead, silver, and zinc). During this

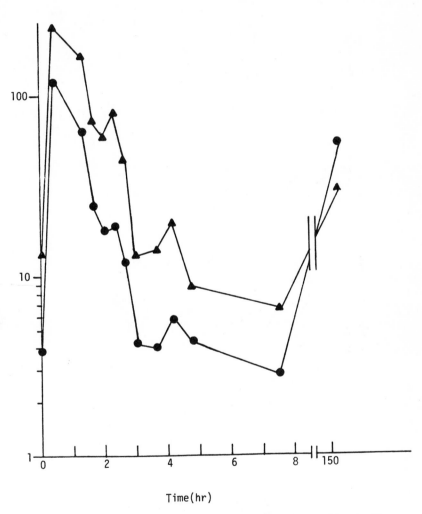

Figure 2. Uptake data for aluminum. (●) Moles of Al/g of cell × 10⁶, (▲) moles of Al/cell × 10⁸.

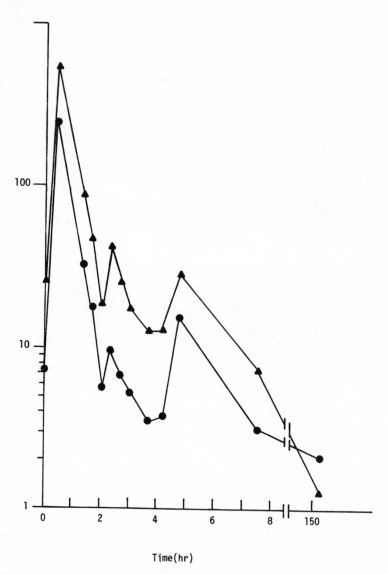

Time(hr)

Figure 3.   Uptake data for cadmium. (▲) *Moles of Cd/cell* × 10²¹, (●) *moles of Cd/g of cell* × 10⁹.

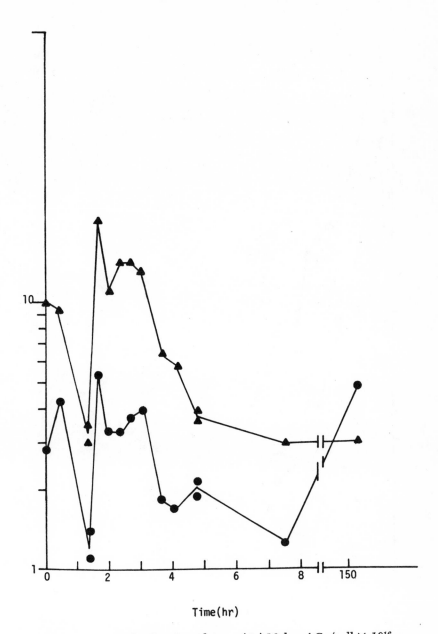

Time(hr)

*Figure 4.   Uptake data for calcium. (▲) Moles of Ca/cell × 10¹⁶,
(●) moles of Ca/g of cell × 10⁴.*

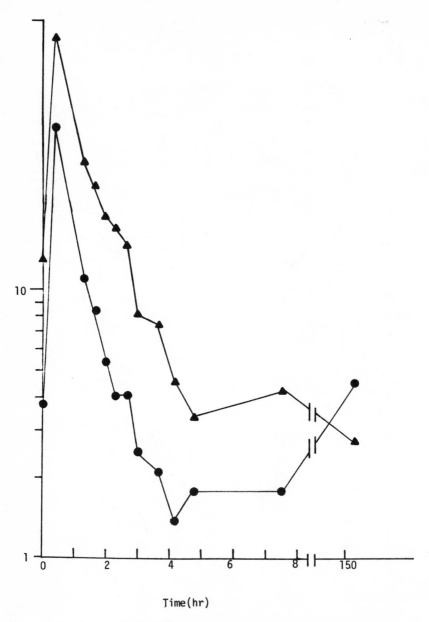

Time (hr)

Figure 5.   Uptake data for chromium. (●) *Moles of Cr/g of cell* × 10⁷, (▲) *moles of Cr/cell* × 10⁹.

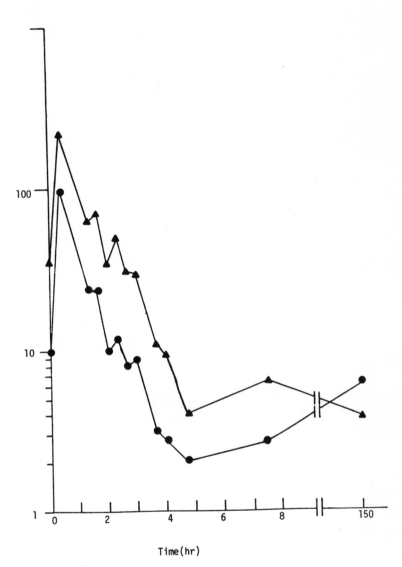

Time(hr)

Figure 6. Uptake data for copper. (▲) *Moles of Cu/cell* × 10^{19}, (●) *moles of Cu/g of cell* × 10^{7}.

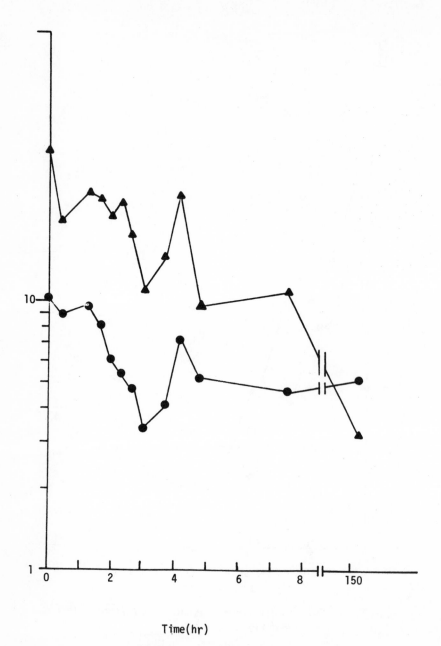

Figure 7. Uptake data for iron. (▲) Moles of Fe/cell $\times$ 10^{18}, (●) moles of Fe/g of cell $\times$ 10^6.

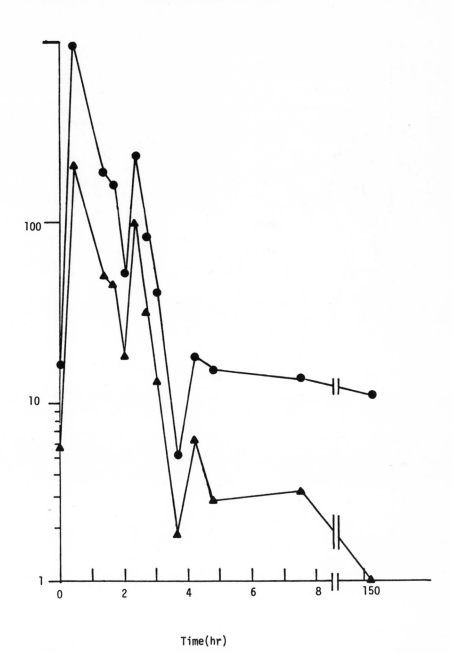

Figure 8. Uptake data for lead. (●) Moles of Pb/g of cell × 19⁹,
(▲) moles of Pb/cell × 10²⁰.

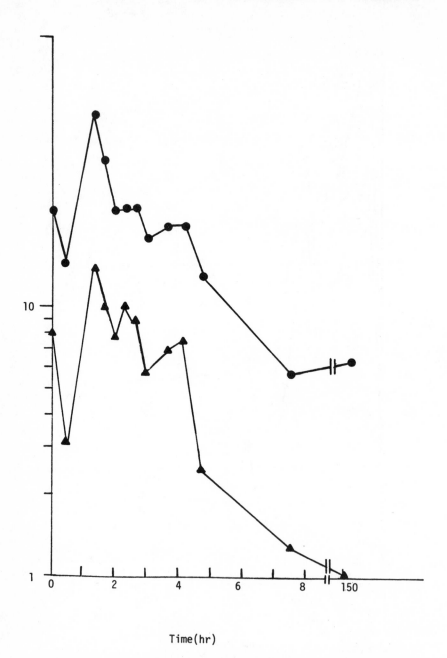

Figure 9.   Uptake data for magnesium. (▲) Moles of Mg/cell
× 10¹⁶, (●) moles of Mg/g of cell × 10⁴.

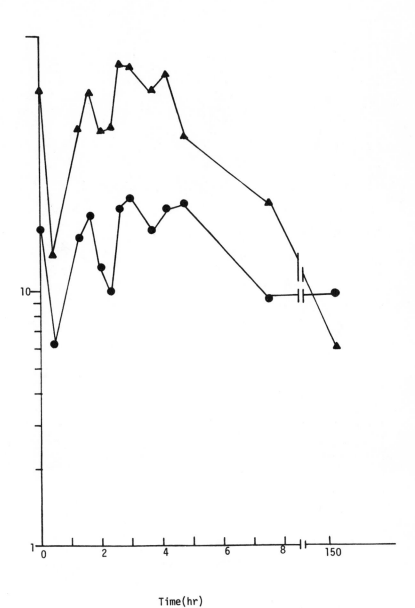

Time(hr)

*Figure 10.   Uptake data for manganese. (▲) Moles of Mn/
cell × 10¹⁹, (●) moles of Mn/g of cell × 10⁷.*

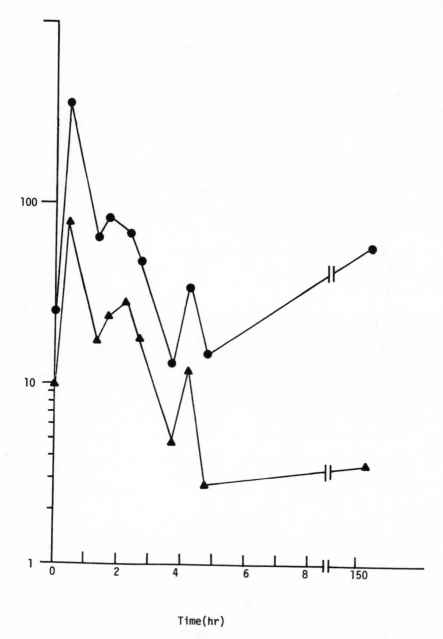

Time(hr)

Figure 11.   Uptake data for silver. (▲) *Moles of Ag/cell* $\times$ *10^{21},* (●) *moles of Ag/g of cell* $\times$ *10^{10}.*

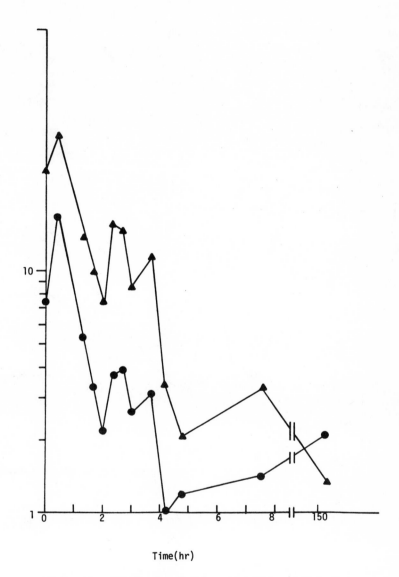

Figure 12.   *Uptake data for zinc.* (▲) *Moles of Zn/cell* × 10^{18}, (●) *moles of Zn/g of cell* × 10^6.

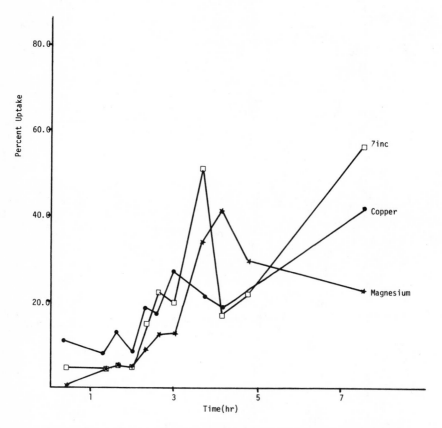

Figure 13. Percent uptake data for copper, magnesium, and zinc

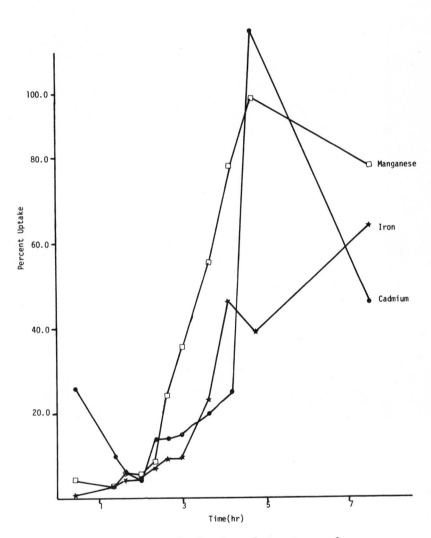

Figure 14.   Percent uptake data for cadmium, iron, and manganese

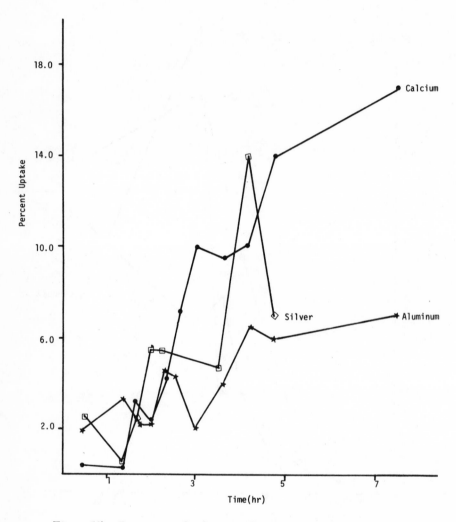

Figure 15.   Percent uptake data for aluminum, calcium, and silver

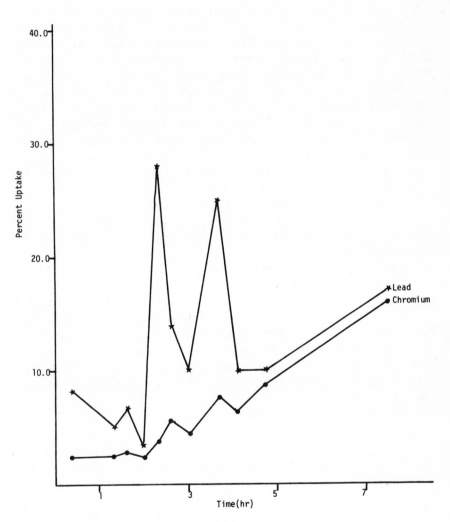

Figure 16.   Percent uptake data for lead and chromium

time the cells are dividing, which means that any metals that increase in concentration are probably involved in a cellular process. Metals that decrease in concentration could be released by the cell or partitioned between the daughter cells. The level of iron remains approximately the same which implies that the uptake is keeping pace with cell division. During exponential growth there is a stage when the dry weight of the cell reaches a maximum (ca. 150 min) and this phenomenon is exhibited as either a maximum or a minimum in the metal uptake data. The data suggest that during this period of exponential growth a change is occurring in the maturation of B. subtilis 168. A second series of maxima and minima for the metal uptake and cell dry weight occurs around 240 min, which suggests that another change is occurring at the end of the exponential growth that probably heralds the onset of the stationary phase. This phenomenon is exhibited to greater or lesser extents by all of the metals monitored. These series of maxima or minima are probably the result of the cells using a different carbon source for growth.

In the stationary phase, if the data representation shown in Figures 2–12 were only considered, it would appear that the metal levels were increasing or decreasing very slightly. This would suggest that once the cell division rate had started to decrease, the metal requirements of the cell decreased or increased more slowly. However, if Figures 13, 14, 15, and 16 are considered, it is apparent that the percentages of the metals available in the growth medium are decreased. It should be stressed at this point that it is possible that not all of the metals found in the TSB before growth are bioavailable, and in fact, the stationary growth phase or the sporulation growth phase occurs when the amount of bioavailable metals decrease to levels that can no longer support exponential growth. This statement is supported by the percent uptake data for copper, zinc, iron, and manganese.

The data shown in Figures 2–12 appear to be contradictory since metal levels appear to increase with respect to the dry weight of the cell and to decrease with respect to the number of moles per cell. This apparent dichotomy can be rationalized easily when one considers that the weight of the spore is much less than the weight of vegetative cell, attributable to the dehydration and to the physical dimensions of the spore. Therefore, this apparent loss in weight is reflected in an apparent increase in metal content in terms of moles of metal per dry weight of cell. However, with the exception of aluminum, which increases substantially in the spore compared with the vegetative cell, and of calcium and silver, whose level remains the same, the levels of all the other metals decrease.

The levels of metals found in vegetative cells were comparable on a dry weight basis for aluminum, calcium, copper, iron, magnesium, manganese, and zinc to previous studies with other species in the *Bacillus* genus (5), and the uptake curves on a dry weight basis for calcium, magnesium, and manganese were similar to studies with *B. subtilis* W23 (*13*). However, no uptake data is available for aluminum, cadmium, chromium, copper, iron, lead, silver, and zinc. While it is possible that most of these metals are required for some function in the maturation cycle of *B. subtilis* 168, it is also probable that cadmium, lead, and silver are not required. No involvement of cadmium, lead, and silver has been reported for any biochemical process, which suggests that these elements are substituting for zinc, calcium, and perhaps sodium, respectively.

However, it is significant that two of the metals that are considered toxic, namely cadmium and lead, are found in the vegetative cells and in the spores without having any apparent effect on the normal maturation of the cell. In conclusion, while no attempt has been made to postulate the roles these 11 metals play in the maturation scheme, this analytical methodology would be suitable for such a study. The requirements of *B. subtilis* 168 for these 11 metals could be confirmed by future studies in which metal deficient growth media were used.

Literature Cited

1. Dulka, J. J., Risby, T. H., *Anal. Chem.* (1976) **48**, 640A.
2. Schleifer, K. H., Kandler, O., *Bacteriol. Rev.* (1972) **36**, 407.
3. Lindberg, A. S., *Annu. Rev. Microbiol.* (1973) **27**, 205.
4. Arber, W., Linn, S., *Annu. Rev. Biochem.* (1969) **38**, 407.
5. Murrell, W. G., *Adv. Microbiol. Physiol.* (1967) **1**, 133.
6. Slepecky, R., Foster, J. W., *J. Bacteriol.* (1959) **78**, 117.
7. Aoki, J. H., Slepecky, R., "Spore Research," A. N. Baker, G. W. Gould, J. Wolf, Eds., p. 93, Academic, New York, 1976.
8. Eisenstadt, E., Fisher, S., Der, C., Silver, S., *J. Bacteriol.* (1973) **113**, 1363.
9. Fisher, S., Buckbaum, L., Toth, K., Eisenstadt, E., Silver, S., *J. Bacteriol.* (1973) **113**, 1373.
10. Scribner, H., Eisenstadt, E., Silver, S., *J. Bacteriol.* (1973) **113**, 1363.
11. Eisenstadt, E., *J. Bacteriol.* (1972) **112**, 264.
12. Silver, S., Toth, K., Scribner, H., *J. Bacteriol.* (1975) **122**, 880.
13. Scribner, H. E., Mogelson, J., Eisenstadt, E., Silver, S., "Spores VI," H. O. Halvorson, R. Hanson, L. L. Campbell, Eds., p. 346, American Society of Microbiology, Washington, D.C., 1972.
14. Bonner, F., Freund, T. S., "Spores V," H. O. Halvorson, R. Hanson, L. L. Campbell, Eds., American Society of Microbiology, Washington, D.C., 1972.
15. Eisenstadt, E., Silver, S., "Spores V," H. O. Halverson, R. Hanson, L. L. Campbell, Eds., p. 443, American Society of Microbiology, Washington, D.C., 1972.
16. Ibid., p. 180.
17. Begnoche, B. C., Risby, T. H., *Anal. Chem.* (1975) **47**, 1041.
18. Gorsuch, T. T., "The Destruction of Organic Matter," Pergamon, London, 1970.

19. Stanier, R. Y., Donboroff, M., Adelberg, E. A., "The Microbial World," Prentice-Hall, New Jersey, 1970.
20. Gould, G. W., Hurst, A., "The Bacterial Spore," Academic, New York, 1969.
21. Curran, H. R., Brunstetter, B. C., Moyers, A. T., *J. Bacteriol.* (1943) **45**, 405.

RECEIVED November 21, 1977.

Role of Zinc in Humans

ANANDA S. PRASAD

Division of Hematology, Department of Medicine, Wayne State University School of Medicine, and Harper Hospital, Detroit, MI 48201, and Veteran's Administration Hospital, Allen Park, MI 48101

In this critical review, the clinical, biochemical, and nutritional effects of zinc in humans have been presented. Deficiency of zinc in man because of nutritional factors and several diseased states has been now recognized. Growth retardation, male hypogonadism, skin changes, poor appetite, mental lethargy, and delayed wound healing are some of the manifestations of chronically zinc-deficient human subjects. In several zinc-deficient patients, dermatological manifestations, diarrhea, alopecia, mental disturbances, and intercurrent infections predominate, and if untreated the condition becomes fatal. Zinc is involved in many biochemical functions and is required for each step of cell cycle in microorganisms. It is essential for DNA synthesis. The activities of many zinc-dependent enzymes have been shown to be affected adversely in zinc-deficient tissues.

Zinc was shown to be essential for microorganisms nearly 100 years ago. In 1934 two groups of workers were able to show convincingly that zinc was required for the growth and well being of rats (*1,2*). In 1955 a disease called parakeratosis of swine was related to zinc deficiency (*3*), and in 1958 it was shown to be essential for the birds (*4*). Clinical manifestations in zinc-deficient animals include growth retardation, testicular atrophy, skin changes, and poor appetite.

In man, deficiency of zinc was suspected to occur for the first time in 1961 (*5*). In 1963, following detailed studies in Egypt, it was shown that nutritional deficiency of zinc did occur in human subjects, thus establishing the essentiality of zinc for man (*6,7*). During the past decade, conditioned deficiency of zinc attributable to many diseased states now has been recognized (*8,9,10*).

0-8412-0416-0/79/33-172-197$08.25/0

Zinc is involved in many biochemical functions. In 1940 Keilin and Mann (11) reported for the first time that carbonic anhydrase was a zinc metalloenzyme. Over the next 20 years, only five additional zinc metalloenzymes were identified, but in the past 15 years, the total number of related enzymes from different species is more than 70 (12). Besides its role in the functions of enzymes, recently it has been shown that zinc may be involved in maintenance of structures of polynucleotides, apoenzymes, and biomembranes (13, 14).

During the past decade or so, rapid advances have been made with respect to the role of zinc in human metabolism. In this review, some of the important aspects of zinc as it relates to human metabolism will be presented.

Nutritional Deficiency of Zinc

In the fall of 1958, a 21-year-old patient at Saadi Hospital, Shiraz, Iran, who looked like a 10-year-old boy, was brought to my attention. In addition to dwarfism, he had severe anemia, hypogonadism, hepatosplenomegaly, rough and dry skin, mental lethargy, and geophagia (5). The nutritional history was interesting in that he ate only bread made of flour. The intake of animal protein was negligible. He also consumed nearly one pound of clay daily. The habit of geophagia (clay eating) is not uncommon in the villages around Shiraz. During a short period of time, we found 10 additional similar cases.

Although there was no evidence for blood loss, the anemia was caused by iron deficiency. We concluded that this may have been attributable to a lack of availability of iron from the bread, greater iron loss attributable to excessive sweating in hot climate, and adverse effect of geophagia on iron absorption. In every case, the anemia completely corrected by administration of oral iron.

It was difficult to explain all of the clinical features solely on the basis of tissue iron deficiency, inasmuch as growth retardation and testicular atrophy are not seen in iron-deficient experimental animals. The possibility that zinc deficiency may have been present was considered. Zinc deficiency was known to produce growth retardation, testicular atrophy, and skin changes in animals. Inasmuch as heavy metals may form insoluble complexes with phosphates, we speculated that some factors responsible for decreased availability of iron in these patients may also adversely affect the availability of zinc.

Subsequently, similar patients were encountered in Egypt (6, 7). Their dietary history was also similar, except that geophagia was not documented. These patients wer studied in detail. These subjects were documented to have zinc deficiency. This conclusion was based on the following: (a) the zinc concentrations in plasma, red cells, and hair were decreased; and (b) radioactive zinc-65 studies revealed that the plasma zinc turnover rate was greater, the 24-hour exchangeable pool was smaller, and the excretion of zinc-65 in stool and urine was less in the patients than in the control subjects.

Further studies in Egypt showed that the rate of growth was greater in patients who received supplemental zinc compared with those who received iron instead or with those who received only animal protein diet consisting of bread, beans, lamb, chicken, eggs, and vegetables. Pubic hair appeared in all cases within 7–12 weeks after zinc supplementation was initiated. Genitalia size became normal, and secondary sexual characteristics developed within 12–24 weeks in all subjects receiving zinc. On the other hand, no such changes were observed in a comparable length of time in the iron-supplemented group or in the group on an animal protein diet. Thus, the growth retardation and gonadal hypofunction in these subjects was related to zinc deficiency.

In 1966 Coble et al. (15) reported a follow-up study of patients with dwarfism and hypogonadism from Kharga Oasis, originally studied by Prasad et al. (16) in 1963. Three years later, a majority of those dwarfs showed an increase in growth and gonadal development without any specific treatment. The plasma zinc concentrations in these patients in 1965 were not altered compared with their levels in 1962. The authors concluded that these cases merely represented examples of delayed maturation. In another paper, Coble et al. (17) demonstrated low plasma zinc levels in normal rural male Egyptians. These results were interpreted to show a lack of relationship between growth and status of zinc in the human body, and the essentiality of zinc in human nutrition was questioned. It is clear from the reports of Coble et al. (15, 17) that those dwarfs from Kharga Oasis, who showed increased growth three years later and hypozincemia, clearly demonstrated delayed maturation and up to the ages of 18–19 failed to attain heights which one would expect to see in Egyptians of an upper socioeconomic level and normal Americans who showed normal plasma zinc levels. The same was true for "normal" rural Egyptian males who showed delayed maturation and smaller ultimate statures compared with their upper socioeconomic counterparts in Cairo. Racially and culturally speaking, the normal subjects from Cairo, belonging to an upper socioeconomic group, are the same as rural "normals." Thus, the so-called rural normals had abnormal growth patterns. Therefore, these observations could be interpreted to show that a low plasma zinc level indeed correlated with slower growth rate and delayed maturation as reported by Coble, Schulert, and Farid (15) and Coble et al. (17). Various degrees of zinc-deficient states may exist in given populations. These cases studied by Prasad et al. (6, 7) were selected from the villages as examples of moderately severe, chronic zinc deficiency. A milder form may manifest itself only by delayed maturation and slower growth rates.

Further studies in Iran clearly demonstrated that zinc is a principal limiting factor in the nutrition of children (18). It was also evident from several studies conducted in the Middle East that the requirement of zinc under different dietary conditions varied widely. For instance, in the studies reported by Prasad et al. (5, 7) and Sandstead et al. (19), 18 mg of supplemental zinc with adequate animal protein and calories diet was sufficient to produce a definite response with respect to growth and gonads, but in other studies when the subjects continued to eat village diet, up to 40 mg of zinc supplement was required to show some growth effect (18).

Halsted et al. (20) published results of their study in a group of 15 men who were rejected at the Iranian Army Induction Center because of "malnutrition." A unique feature was that all were 19 or 20 years of age. Their clinical features were similar to those of zinc-deficient dwarfs reported earlier by Prasad et al. (6, 7). They were studied for 6 to 12 months. One group was given a well-balanced nutritious diet containing ample animal protein plus a placebo capsule. A second group was given the same diet plus the diet alone without additional medication for six months, followed by the diet plus zinc for another six-month period.

The development in subjects receiving the diet alone was slow, and the effect on height increment and onset of sexual function was strikingly enhanced in those receiving zinc. The zinc-supplemented boys gained considerably more height than those receiving ample protein diet alone. The zinc-supplemented subjects showed evidence of early onset of sexual function, as defined by nocturnal emission in males and menarche in females. The two women described in this report were from hospital clinic and represented the first cases of dwarfism in females because of zinc deficiency (20).

A brief mention should be made regarding the prevalence of zinc deficiency in human populations throughout the world. Clinical pictures similar to those reported by us in zinc-deficient dwarfs have been observed in many countries such as Turkey, Portugal, and Morocco (21). Also, zinc deficiency should be prevalent in other countries where primarily cereal proteins are consumed by the population. Clinically, perhaps it is not very difficult to recognize extreme examples of zinc-deficient dwarfs in a given population, but it is the marginally deficient subjects who would present great difficulties, and only future studies can provide insight into this problem. It is becoming clear now that not only nutritional deficiency but also conditioned deficiency of zinc may complicate many diseased states.

Research on the nutritional status of zinc in infants and young children has been very limited. However, the importance of adequate zinc nutrition in the young is apparent from data on other mammals. Dietary zinc requirements are relatively high in the young as compared with those of mature animals of the same species, and the effects of dietary insufficiency are particularly severe. Furthermore, several of the major features of zinc deficiency, such as growth retardation and impaired learning ability, are peculiar to the young animal.

A clinical syndrome similar to that of "adolescent nutritional dwarfism" has been identified in younger children in Iran, though failure of sexual maturation is not evident prior to adolescence. This syndrome is most common in small, rural communities in which there is also a high incidence of adolescent nutritional dwarfism.

Plasma zinc levels have been measured in infants and young children suffering from kwashiorkor in Cairo, Pretoria, Cape Town, and Hyderabad (9). At the time of hospital admission, levels were very low in all four locations, but the hypozincemia could be attributed at least in part to hypoalbuminemia. During the subsequent eight weeks, plasma zinc levels increased, and in Cape Town and Hyderabad, reached normal control levels. However, in Cairo and Pretoria plasma zinc levels remained significantly below normal at a time of "clinical cure," when total serum protein and serum albumin levels were normal. This persistent hypozincemia suggests that zinc deficiency is associated with kwashiorkor in some geographical locations. The clinical significance of this deficiency has not been defined but may, for example, have contributed to the growth failure and to the incidence of skin ulceration. After recovery from kwashiorkor, children in some areas of the world, including Egypt, are likely to receive a diet inadequate in zinc, and a deficiency of this nutrient may persist indefinitely.

In Cape Town the zinc concentration in the liver but not in the brain, heart, or muscle of children dying from kwashiorkor was significantly lower than normal. Plasma zinc levels were also low in marasmic infants.

Nutritional Zinc Deficiency in Children in the United States. In 1972 a number of Denver children were reported to have evidence of symptomatic zinc deficiency (22). These children were identified as a result of a survey of trace element concentrations in the hair of apparently normal children from middle- and upper-income families. Ten of a group of 132 children, 4–16 years, had hair zinc concentrations less than 70 ppm, or more than three SD below the normal adult mean. This was an unexpected finding in normal children who had had no recent or chronic illness, and consequently additional data were obtained.

Eight of these ten children were found to have heights at or below the 10th percentile on the Iowa Growth Charts, though the children included in this study were not preselected according to height. There was no apparent cause of the relatively poor growth, which could not be explained on a familial basis. Growth retardation is one of the earliest manifestations of zinc deficiency in the young animal, and the correlation between the low hair zinc levels and low growth percentiles in these children suggested a causal relationship.

Anorexia is another prominent and early feature of zinc deficiency in animals, and most of the children with low hair zinc levels in the original Denver study also had a history of poor appetite. In particular, the consumption of meats was very limited despite access to larger quantities. As animal products are the best source of available zinc, it is quite possible that the dietary zinc intake of these children was inadequate.

It has been calculated recently that substantial sections of the population of the United States are at risk from suboptimal zinc nutrition (23). Those at particular risk from a deficiency of this metal include subjects whose zinc requirements are relatively high, for example, times of rapid growth and times when people subsist on low-income diets.

Zinc Nutrition and Deficiency in Infants. Hair and plasma zinc levels are exceptionally low in infants in the United States compared with other age groups including the neonate, older children, and adults (24). In the Denver survey of hair trace-element concentrations discussed above, there was a mean hair zinc level of only 74 ± 8 ppm for the 26 apparently normal infants, 3–12 months, compared with 174 ± 8 ppm for neonates and 180 ± 4 ppm for adults. Fifty-four percent of these infants from middle- and upper-income families had hair zinc levels less than 70 ppm. Exceptionally low hair zinc levels also have been found in infants residing in Dayton, Ohio (25, 26). It appears unlikely that these low levels can be regarded as entirely normal for this age, as levels are not equally low in other countries where comparable data have been obtained. For example, in Thailand the mean hair zinc level of 15 infants from upper economic homes was 202 ± 26 ppm; no source of external contamination could be identified in Bangkok to account for these higher levels. Hair zinc concentrations of adults in Thailand were closely comparable with those of Denver adults, but age-related differences in the younger children as well as in infants were very different from those in Denver. These discrepancies indicate that the low levels in Denver infants cannot necessarily be accepted as physiological.

Similar considerations apply equally to the low plasma zinc levels reported for infants in this country. Plasma concentrations in infants in Sweden and Germany have been reported to be no lower than those for adults in these countries (27, 28)

Several factors, including difficulty in achieving positive zinc balance in early postnatal life and a "dilutional" effect of rapid growth, may contribute to zinc depletion in infants (29). A unique factor in the United States which may contribute to zinc deficiency is the low concentration of this metal in certain popular infant milk formulae. Formulae in which the protein content of original cow's milk is reduced to levels comparable with those of human milk have a parallel reduction in zinc content. Unless subsequently supplemented with zinc, these formulae typically have a zinc concentration of less than 2 mg/L, or approximately one-half that of cow's milk and of human milk during the first two months of lactation. Zinc supplementation of all of these formulae has not been routine, though it is likely to become a universal practice in the near future following the recent recommendations of the Food and Nutrition Board of the National Academy of Sciences with respect to zinc intake in infants. Indirect support for the possible role of these formulae in the low hair zinc levels of U.S. infants has been derived from measurements of hair zinc concentrations in English formula-fed infants. The mean hair-zinc level of these infants was 124 ± 12 ppm which, though lower than adults, was 68% higher than the Denver mean ($p < 0.005$).

The majority of infants with low levels of zinc in plasma and hair have not had any detectable signs of zinc deficiency. However, it appears that those at the lower end of a spectrum of zinc depletion, as manifested by low levels of hair and plasma zinc, and perhaps those who remain moderately depleted for a prolonged period of time, do develop symptomatic zinc deficiency. In the original Denver study, 8 of 93 infants and children aged less than four years had hair zinc levels less than 30 ppm, and six of these manifested declining growth percentiles and poor appetite (29). It was noted also that the high percentiles of the older children with hair zinc levels less than 70 ppm first declined during infancy; thus, if their poor growth resulted from an insufficiency of this nutrient, the latter must have commenced during infancy. It is conceivable that once a deficiency state has been established, the resulting anorexia may tend to perpetuate the deficiency state. Experience with a number of infants who had low levels of zinc in plasma and hair and who have responded favorably to zinc supplementation, indicates that some cases of failure to thrive in infancy may be caused by zinc deficiency. Anorexia has been a prominent feature in these infants, and one case manifested a bizarre form of pica which improved dramatically following zinc supplementation (29). Currently, however, there is a lack of controlled studies of dietary zinc supplementation in these infants.

Other Causes of Zinc Deficiency

Alcohol. Alcohol induces hyperzincuria (8, 9). The mechanism is unknown. A direct effect of alcohol on renal tubular epithelium may be responsible for hyperzincuria. Acute ingestion of alcohol did not induce zincuria (30) in some experiments, however, Gudbjarnason and Prasad (31) reported increased urinary zinc excretion following alcohol intake. This effect was evident when complete urine collection was analyzed for zinc during first three-hour and second three-hour periods, following ingestion of six ounces of chilled vodka.

The serum zinc level of the alcoholic subjects tends to be lower in comparison with the controls. An absolute increase in renal clearance of zinc in the alcoholics demonstrable at both normal level and low serum zinc concentration has been observed (32). Thus the measurement of renal clearance of zinc may be clinically used for etiological classification of chronic liver disease attributable to alcohol in different cases.

Excessive ingestion of alcohol may lead to severe deficiency of zinc. In one case, acquired zinc deficiency attributable to alcohol was characterized by mental disturbances, widespread eczema craquele, hair loss, steatorrhea, and dysproteinemia with edema (33). Therapy with zinc reversed these manifestations. A similar clinical syndrome has been seen among Ugandian blacks addicted to banana gin.

Liver Disease. Valee et al. (34) initially described the abnormal zinc metabolism that occurs in patients with cirrhosis of the liver. These investigators demonstrated that patients with cirrhosis had low serum zinc, diminished hepatic zinc, and paradoxically, hyperzincuria. These observations led them to suggest that zinc deficiency in the alcoholic cirrhotic patient may be a conditioned deficiency which was somehow related to alcohol ingestion. These observations have been confirmed by other investigators.

Recently, Morrison et al. (35) reported abnormal dark adaptation in six stable alcoholic cirrhotics who also had low serum zinc levels. Zinc administration to these patients resulted in improvement of dark adaptation. This is an interesting clinical observation but needs confirmation. Saldeen and Brunk (36) have shown that parenteral zinc salts protect rat liver from damage by carbon tetrachloride. These studies suggest that zinc exerts a protective effect on the liver.

It is likely that some of the clinical features of cirrhosis of the liver such as loss of body hair, testicular hypofunction, poor appetite, mental lethargy, difficulty in wound healing, and night blindness may indeed be related to the secondary zinc-deficient state in this disease. In the future, careful clinical trials with zinc supplementation must be carried out to determine whether or not zinc is beneficial to patients with chronic liver disease.

Gastrointestinal Disorders. Zinc deficiency has been reported in patients with steatorrhea. In an alkaline environment, zinc would be expected to form insoluble complexes with fat and phosphate analogous to those formed by calcium and magnesium. Thus, fat malabsorption attributable to any cause should result in an increased loss of zinc in the stool.

Exudation of large amounts of zinc-protein complexes into the intestinal lumen also may contribute to the decrease in plasma zinc concentrations which occur in patients with inflamamtory disease of the bowel. It seems likely that protein losing enteropathy because of other causes also may impair zinc homeostasis. Another potential cause of negative zinc balance is a massive loss of intestinal secretions.

Renal Disease. The potential causes of conditioned deficiency of zinc in patients with renal disease include proteinuria and failure of tubular reabsorption. In the former instance, the loss of zinc–protein complexes across the glomerulus is the mechanism. In the latter an impairment in the metabolic machinery of tubular reabsorption attributable to a genetic abnormality or to toxic substances would result in zinc loss. While low plasma zinc concentrations have been described in patients with massive proteinuria, no reports of low plasma levels of zinc in patients with tubular reabsorption defects have appeared in literature (9).

In patients with renal failure, the occurrence of conditioned zinc deficiency may be the result of a mixture of factors, which at present are ill defined. If 1,25-dihydroxycholecalciferol plays a role in the intestinal absorption of zinc, an impairment in its formation by the diseased kidney would be expected to result in malabsorption of zinc. It seems likely that plasma and soft tissue concentrations of zinc may be "protected" in some individuals with renal failure by the dissolution of bone which occurs as a result of increased parathyroid activity in response to low serum calcium. In experimental animals, calcium deficiency has been shown to cause release of zinc from bone. In some patients who are successfully treated for hyperphosphatemia and hypocalcemia, the plama zinc concentration may be expected to decline because of the deposition of zinc along with calcium in bone. Thus, in the latter group in particular, a diet low in protein and high in refined cereal products and fat would be expected to contribute to a conditioned deficiency of zinc. Such a diet would be low in zinc. The patients reported by Mansouri et al. (37), who were treated with a diet containing 20–30 g of protein daily and who had low plasma concentrations of zinc, appear to represent such a clinical instance. Presumably the patients of Halsted and Smith (38) were similarly restricted in dietary protein. In other patients with renal failure whose dietary protein was not restricted, plasma zinc concentration were not decreased. Patients on dialysis had even higher levels, particularly following dialysis. Apparently zinc deficiency may not be a problem in patients on dialysis if their dietary consumption of protein is not restricted.

Neoplastic Diseases. The occurrence of conditioned deficiency of zinc in patients with neoplastic diseases will obviously depend upon the nature of the neoplasm. Anorexia and starvation, plus avoidance of foods rich in available zinc, are probably important conditioning factors. An increased excretion of zinc subsequent to its mobilization by leukocyte endogenous mediator (LEM) in response to tissue necrosis may be another factor.

Burns and Skin Disorders. The causes of zinc deficiency in patients with burns include losses in exudates. Starvation of patients with burns is a well-recognized cause of morbidity and mortality. The contribution of conditioned zinc deficiency to the morbidity of burned patients is not defined. Limited studies indicate that epithelialization of burns can be improved by treatment with zinc. Such a finding is consistent with the beneficial effect of zinc on the treatment of leg ulcers and the well-defined requirement of zinc for collagen synthesis (39, 40, 41, 42).

In psoriasis, the loss of large numbers of skin cells possibly may result in zinc depletion. The skin contains approximately 20% of the body zinc. Thus, if the loss of epithelial cells is great enough, it is conceivable that the massive formation of new cells by the skin could lead to conditioned deficiency. Low levels of plasma zinc have been reported in some patients with extensive psoriasis (43, 44), however, others have not been able to confirm these findings (45, 46).

Impaired Wound Healing in Chronically Diseased Subjects. In 1966 Pories and collaborator (47, 48) reported that oral administration of ZnSO$_4$ to military personnel with marsupialized pilonidal sinuses was attended by a twofold increase in the rate of reepithelialization. The authors' conclusion that ZnSO$_4$ can promote the healing of cutaneous sores and wounds has been subject of controversy during the past several years. Clinical investigation by Cohen (49), Husain (39), Greaves and Skillen (50), and Serjeant et al. (40), have substantiated the beneficial effects of ZnSO$_4$ on wound healing whereas studies by Barcia (51), Myers and Cherry (52), Clayton (53), and Greaves and Ive (54) have failed to demonstrate any therapeutic benefit. Hallbook and Lanner (55) found that the reepithelialization rate of venous leg ulcers was enhanced by ZnSO$_4$ in patients who initially had diminished concentrations of serum zinc, but they did not find any benefit in patients whose initial measurements of serum zinc were within the normal range.

Studies in experimental animals have demonstrated that: (a) healing of incised wounds is impaired in rats with dietary zinc deficiency; (b) collagen and noncollagen proteins are reduced in skin and connective tissues from rats with dietary zinc deficiency; (c) zinc supplementation does not augment wound healing in normal rats; and (d) zinc supplementation does augment wound healing in chronically ill rats (56, 57). These data provide evidence that zinc supplementation may promote wound healing in zinc-deficient patients.

Parasitic Infestations. Blood loss attributable to parasitic diseases may contribute to conditioned deficiency of zinc. Such appears to have been the case in the zinc responsive "dwarfs" reported from Egypt (6, 7). As red blood cells contain 12–14 μg of Zn/mL, infections with hookworm and/or schistosomiasis which are severe enough to cause iron deficiency will probably contribute to the occurrence of zinc deficiency.

Iatrogenic Causes. Possible iatrogenic causes of conditioned deficiency of zinc include use of antimetabolites and antianabolic drugs. Treatment with some of these drugs makes patients feel ill. They become anoretic and may starve. With catabolism of body mass, urinary excretion of zinc is increased. Commonly used intravenous fluids are relatively zinc free. Thus, under usual circumstances, a negative zinc balance should occur in patients who are given antimetabolites, antianabolic agents, or prolonged intravenous therapy.

Failure to include zinc in fluids for total parenteral nutrition (TPN) is another example of iatrogenically induced, conditioned deficiency of zinc. A decline of plasma zinc has been observed in several patients given TPN fluids containing less than 1.25 mg of zinc daily (9, 58). The decrease which occurred was probably not attributable to urinary losses. Some patients on TPN have developed a clinical picture resembling acrodermatitis enteropathica.

Zinc deficiency occurring in patients following penicillamine therapy for Wilson's disease has been reported recently (59). The manifestations consisted of parakeratosis, "dead" hair and alopecia, keratitis, and centrocecal scotoma. The clinical features were similar to those of acrodermatitis enteropathica. Following supplementation with zinc, several clinical manifestations were reversed.

Diabetes. Some patients with diabetes mellitus have been found to have increased urinary losses of zinc (60). The mechanism is unknown. Presumably, some of them may become zinc deficient though in general, the plasma zinc is not affected. Perhaps the failure of some diabetics to heal ulcers of their feet and elsewhere is related to zinc deficiency. Healing of such ulcers in diabetes has been reported subsquent to zinc therapy (61).

Collagen Diseases. In patients with inflammation such as rheumatoid arthritis, lupus erythematosus, infection, or injury, two factors may lead to zinc deficiency. The loss of zinc from catabolized tissue (62) and mobilization of zinc by leucocytic endogenous mediator (LEM) (63) to the liver and its subsequent excretion in the urine may account for conditioned zinc deficiency in such cases. Recently, a beneficial effect of zinc therapy in patients with rheumatoid arthritis has been reported (*64*).

Pregnancy. Plasma concentration of zinc decreases in human pregnancy (*65*). Presumably, the decrease reflects in part the uptake of zinc by the fetus and other products of conception. It has been estimated that the pregnant woman must retain approximately 750 μg of Zn/day for growth of the products of conception during the last two thirds of pregnancy. Thus, when zinc deficiency occurs in pregnancy, a conditioning factor is the demand of the fetus for zinc. Studies in the rat suggest that the placenta actively provides zinc to the fetus (*66*). If the diet of the pregnant woman does not include liberal amounts of animal protein, the likelihood of conditioned deficiency of zinc is increased, as zinc is probably less available from food derived from grains and other plants. The possible importance of zinc deficiency in human pregnancy is implied by the observations reported by Hurley et al. (*67*), Caldwell et al. (*68*), and Halas et al. (*69*). Zinc deficiency in pregnant rats was shown to cause fetal abnormalities, behavioral impairment in the offspring, and difficulty in parturition in the mother (*67*).

Caldwell et al. (*68*) were the first to show that in both prenatal and postnatal nutrition even a mild zinc deficiency in rats had profound influence on behavior potential despite an apparently adequate protein level in the diet. Recently, it has been observed that zinc deficiency in fetal and suckling rats results in adverse biochemical effects in the brain (*69*). The adverse effects on the brain of the suckling rats are apparently not readily reversible, as behavioral testing of nutritionally rehabilitated 60–80-day-old male rats has shown that they performed poorly on a Tolman–Honzik Maze (*70*) when compared with pair-fed and ad libitum-fed control rats. These findings suggested that zinc deficiency during the critical developmental period of the rat brain induces poorly reversible abnormalities which are manifested by impaired behavioral development.

Hurley and co-workers (*67*) have shown that short-term depletion of zinc in maternal rats results in a variety of congenital anomalies in the offspring. In view of the important role of zinc in nucleic acid metabolism, Hurley and Schroeder (*71*) have proposed that impaired deoxyribonucleic acid (DNA) synthesis in zinc-deprived embryos prolongs the mitotic cycle and reduces the number of normal neural cells, leading to malformations of the central nervous system. It is tempting to speculate that the exceptionally high rates of congenital malformations of the central nervous system, as reported from the Middle East (*72*), might be caused by maternal zinc deficiency.

Plasma zinc is known to decrease following use of oral contraceptive agents (73, 74). Our recent data indicate that whereas the plasma zinc may decline, the zinc content of the red blood cells increases as a result of oral contraceptive agents administration. This phenomenon may merely mean a redistribution of zinc from the plasma pool to the red cells. Alternatively, oral contraceptive agents may enhance carbonic anhydrase (a zinc metalloenzyme) synthesis, thus increasing the red cell zinc content.

Genetic Disorders. SICKLE CELL DISEASE. Recently deficiency of zinc in sickle cell disease has been recognized (75). Certain clinical features are common to some sickle-cell-anemia patients and zinc-deficient patients, the latter as reported from the Middle East. These symptoms include delayed onset of puberty and hypogonadism in the males, characterized by decreased facial, pubic, and axillary hair, short stature, and low body weight, rough skin, and poor appetite. Inasmuch as zinc is an important constituent of erythrocytes, it appeared possible that long-continued hemolysis in patients with sickle cell disease might lead to a zinc-deficient state, which could account for some of the clinical manifestations mentioned above. Delayed healing of leg ulcers and the reported beneficial effect of zinc therapy on leg ulcers in sickle-cell-anemia patients would also appear to be consistent with the above hypothesis.

In a study reported by Prasad et al. (75), zinc in plasma, erythrocytes, and hair was decreased and urinary zinc excretion was increased in sickle-cell-anemia patents as compared with controls. Erythrocyte zinc and daily urinary zinc excretion were inversely correlated in the anemia patients ($r = -0.71$, $p < 0.05$), suggesting that hyperzincuria may have caused zinc deficiency in these patients. Carbonic anhydrase, a zinc metalloenzyme, correlated significantly with erythrocyte zinc ($r = +.94$, $p < .001$). Plasma ribonuclease (RNase) activity was significantly greater in anemia subjects than in controls, consistent with the hypothesis that sickle-cell-anemia patients were zinc deficient. Zinc sulfate (660 mg per day) was administered orally to seven men and two women with sickle cell disease. Two 17-year-old males gained 10 cm in height during 18 months of therapy. All but one patient gained weight (10). Five of the males showed increased growth of pubic, axillary, facial, and body hair, and in one male a leg ulcer healed in six weeks on zinc, and in two other males some benefits of zinc therapy on healing of ulcers was noted (10).

These data show that some sickle cell anemia patients are zinc deficient, although the severity varied considerably from one patient to another in this group. In spite of the tissue zinc depletion in sickle-cell-anemia patients, the mean urinary excretion of zinc was higher than in the controls. This may have been directly a result of increased filtration of zinc by the glomeruli, owing to continued hemolysis, or there may have been a defect in tubular reabsorption of zinc somehow related to sickle cell anemia, a possibility that cannot be excluded at present. Continued hyperzincuria may have been responsible for tissue depletion of zinc as suggested by a significant negative correlation between values for 24-hour urinary zinc excretion and erythrocyte zinc. At this stage, however, one cannot rule out additional factors such as predominant dietary use of cereal protein and other nutritional factors that affect zinc availability adversely, thus accounting for zinc deficiency. Further work is warranted for proper elucidation of the pathogenesis of zinc deficiency in sickle cell anemia.

Recent studies have demonstrated a potential beneficial effect of zinc on the sickling process, in vitro, mediated by its effect on the oxygen dissociation curve (76, 77) and the erythrocyte membrane (78). Zinc-bound hemoglobin should be beneficial to sickle-cell-disease patients inasmuch as this hemoglobin has increased oxygen affinity and S hemoglobin sickles only when it is deoxygenated. Whether or not this zinc effect on S hemoglobin can be achieved in vivo remains to be seen. It is conceivable that the zinc-binding residue could be one involved with the α_1–β_2 contact points which might result in decreasing the stability of the contact and favor the oxy form of hemoglobin. More definitive studies, such as x-ray diffraction, are needed before the exact zinc-binding residue and oxygen-affinity mechanism is defined with certainty.

It has been proposed (a) that hemoglobin binding to the inside of the sickle, red-cell membrane is an important feature of the development of the so-called irreversibly sickled cell; (b) that calcium facilitates the binding of hemoglobin particles; (c) that binding of hemoglobin particles to the membrane increases its stiffness and rigidity, perhaps by cross-linking mechanisms; and (d) that zinc protects against these changes by partially blocking the binding of hemoglobin to the membrane either directly or by an interaction with calcium, and as a result less hemoglobin is bound to the membrane in the presence of zinc and the deformability of the cell is thereby protected (78).

In limited uncontrolled studies, zinc appears to have been effective in decreasing symptoms and crises of sickle-cell-anemia patients. The therapeutic rationale is based on effects of zinc on the red cell membrane by which it decreases hemoglobin and calcium binding and improves deformability, which may result in decreased trapping of sickle cells in the capillaries where pain cycle is normally initiated. Undoubtedly more thorough evaluation of zinc therapy in sickle cell disease is needed in the future.

ACRODERMATITIS ENTEROPATHICA. Acrodermatitis enteropathica was described in 1943 by Danbolt and Closs (*79*) and the clinical and pathological features have been delineated by numerous investigators (*80*). In brief, acrodermatitis enteropathica is a lethal, autosomal, recessive trait which usually occurs in infants of Italian, Armenian, or Iranian lineage. The disease is not present at birth, but typically develops in the early months of life, soon after weaning from breast feeding. Dermatological manifestations include progressive bullous-pustular dermatitis of the extremities and of the oral, anal, and genital areas, combined with paronychia and generalized alopecia. Infection with *Candida albicans* is a frequent complication. Ophthalmic signs may include blepharitis, conjunctivitis, photophobia, and corneal opacities. Gastrointestinal disturbances are usually severe, including chronic diarrhea, malabsorption, steatorrhea, and lactose intolerance. Neuropsychiatric signs include irritability, emotional disorders, tremor, and occasional cerebellar ataxia. The patients generally have retarded growth and hypogonadism. Prior to the serendipitous discovery of diiodohydroxyquinolone therapy in 1953 by Dillaha and co-workers (*81*), patients with acrodermatitis enteropathica invariably died from cachexia, usually with terminal respiratory infection. Although diiodohydroxyquinolone has been used successfully for the therapy of this condition for 20 years, the mechanism of drug action has never been elucidated. It now seems possible that the efficacy of diodohydroxyquinolone might be related to the formation of an absorbable zinc chelate, inasmuch as diiodohydroxyquinolone is a derivative of 8-hydroxyquinolone, a chelating agent (*82*).

In 1973, Barnes and Moynahan (*83, 84*) studied a 2-year-old girl with severe acrodermatitis enteropathica who was being treated with diiodohydroxyquinolone and a lactose-deficient synthetic diet. The clinical response to this therapy was not satisfactory and the physicians sought to identify contributory factors. They found that the concentration of zinc in the patient's serum was profoundly reduced and, therefore, they administered oral $ZnSO_4$. The skin lesions and gastrointestinal symptoms cleared completely, and the patient was discharged from the hospital. When $ZnSO_4$ was inadvertently omitted from the child's regimen, she suffered a relapse which promptly responded when oral $ZnSO_4$ was reinstituted. In their initial reports, Barnes and Moynahan (*83, 84*) attributed zinc deficiency in this patient to the synthetic diet.

It was appreciated later that zinc might be fundamental to the pathogenesis of this rare inherited disorder and that the clinical improvement reflected improvement in zinc status. Support for zinc deficiency hypothesis came from the observation that a close resemblance between the symptoms of zinc deficiency in animals and man as reported earlier (*85*) and subjects with acrodermatitis enteropathica existed particularly with respect to skin lesions, growth pattern, and gastrointestinal symptoms.

Zinc supplementation to these patients led to complete clearance of skin lesions and restoration of normal bowel function, which had previously resisted various dietary and drug regimens. This original observation was quickly confirmed in other cases with equally good results. The underlying mechanism of the zinc deficiency in these patients is most likely attributable to malabsorption. The cause of poor absorption is obscure, but an abnormality of Paneth's cells may be involved. These observations should provide a great stimulus to all interested in zinc metabolism to look for manifestations of zinc deficiency in other disorders, either natural or induced, in both children and adults.

MISCELLANEOUS GENETIC DISORDERS. Low levels of plasma zinc have been noted in patients with mongolism (38). The mechanism is unknown. Congenital hypolasia of the thymus gland in cattle may be an example of zinc deficiency on genetic basis (86). It is unknown whether thymus hypoplasia in man is in some way related to zinc deficiency.

Experimental Production of Zinc Deficiency in Man

Although the role of zinc in human subjects has been now defined and its deficiency recognized in several clinical conditions, these examples are not representative of a pure zinc-deficient state in man. It was, therefore, considered desirable to develop a human model which would allow a study of the effects of a mild zinc-deficient state in man and also would provide sensitive parameters which could be used clinically for diagnosing zinc deficiency. Recently marginal deficiency of zinc by dietary means has been produced successfully in human volunteers and changes in several zinc-dependent parameters have been documented (87).

Four male volunteers (1, 2, 3, 4) participated in this study. Their ages ranged from 55 to 65 years. The physical examination revealed no abnormal features and their zinc status was within normal limits.

A semipurified diet based on textured soy protein purchased from General Mills Co., Minneapolis, Minn. (Bontrae Products) and Worthington Foods Co., Division of Miles Laboratory, Elkart, Indiana, was developed for this study. Soy protein isolate, which was used as soy flour in the baked goods, was purchased from General Biochemicals (Teklad Mills, Chagrin Falls, Ohio). The texturized soy meals used were hamburger granules, chicken slices, turkey slices, and chicken chunks. The texturized soy protein and soy protein isolate were washed twice with ethylenediaminetetraacetate (EDTA), then rinsed three times with deionized water, boiled for 30 minutes, and kept frozen until ready to be used. The diet was supplemented with vitamins, mineral mix (except zinc), and protein supplement in order to meet the recommended dietary requirements. The daily intake of zinc was 2.7 mg for patients 1 and 2 and 3.5 mg for patients 3 and 4.

The patients were kept under strict metabolic conditions. The first group of subjects (1 and 2), received hospital diet for two weeks, and then they received the experimental diet with 10 mg of supplemental zinc (as zinc acetate) orally for six weeks. Following this, they were given only experimental diet (daily zinc intake of 2.7 mg) for 24 weeks. At the end of this phase, while continuing the experimental diet the two subjects received 30 mg of zinc (as zinc acetate) supplement daily, orally for 12 weeks. Finally these were maintained on hospital diet with total daily intake of 10 mg zinc plus 30 mg of oral zinc (as zinc acetate) supplement for eight weeks. The hospital diet provided the same amount of calories and protein as the experimental diet. Thus these two subjects were observed for a total period of 52 weeks.

The second group of subjects (3 and 4) received hospital diet (10 mg of zinc intake daily) for three weeks, followed by the experimental diet with 30 mg of oral zinc supplement (as zinc acetate) for five weeks. Following this they were given only experimental diet (3.5 mg of zinc intake daily) for 40 weeks. The repletion phase was begun by giving 30 mg of zinc (as zinc acetate) orally while maintaining the same experimental diet and continued for a total period of eight weeks, at the end of which the hospital diet replaced the experimental diet. Oral zinc supplement (30 mg as zinc acetate) was continued along with the hospital diet for a period of eight weeks. Altogether these two subjects were observed for 64 weeks.

The body weight decreased in all four subjects as a result of dietary zinc restriction and the weight loss was more pronounced in the first group of subjects as compared with the second group. The body weight correlated highly with the subscapular thickness in the two subjects in whom these data were obtained. An approximate calculation revealed that the weight loss could be accounted for as follows: 50% fat, 30% water, and 20% other.

In the first group of two subjects, during the zinc-restriction phase (2.7 mg zinc daily intake), the apparent negative balance for zinc ranged from 1 to 4 mg per day whereas in the second group of subjects, the apparent negative balance for zinc was 1 to 2 mg per day. Following supplementation with 30 mg of zinc, the positive zinc balance ranged from 11 to 22 mg per day in the first group of two subjects. In the second group of subjects (3 and 4), during the baseline period when the daily zinc intake was 33.5 mg daily, the positive balance for zinc was 3–4 mg daily. On the other hand, these subjects, on the same level of zinc intake (33.5

mg daily) following zinc depletion phase, showed a positive zinc balance of 14–16 mg daily.

The plasma zinc decreased significantly in all four subjects as a result of zinc restriction and increased following supplementation with zinc. The changes were more marked for patients on 2.7 mg of daily zinc intake as compared with those on 3.5 mg daily zinc intake. The red cell zinc decreased significantly in the first group of two subjects (1 and 2) although the decrease was not evident until 12 weeks on restricted zinc intake. In the second group of subjects, although the red cell zinc did not decrease significantly during zinc-restricted period, it showed a marked increase following zinc supplementation. The leucocyte zinc decreased significantly in the second group of subjects as a result of zinc restriction, in whom this parameter was measured.

Plasma alkaline phosphatase was monitored carefully in the second group of subjects. In both cases, the activity slowly declined as a result of zinc restriction, and following supplementation with zinc, the activity nearly doubled in eight weeks. In all four subjects, the activity of plasma ribonuclease was almost twice as great during the zinc-restricted period as compared with the zinc supplemented phase. Surprisingly, it was observed that plasma ammonia levels increased during zinc restriction and decreased following zinc supplementation in the second group of two subjects in whom this was monitored.

Urinary excretion of zinc decreased in three out of four subjects as a result of zinc restriction. In one case, the decrease in urinary zinc excretion was not seen. This was attributable to the diuretic therapy which he received for mild hypertension during the study.

Total protein, total collagen, ribonucleic acid (RNA), DNA, and deoxythymidine kinase activity were measured in the connective tissue obtained after implantation subcutaneously of a small sponge during zinc-restriction and zinc-supplementation phases in the first group of two subjects. A marked increase in the total protein, total collagen, and RNA/DNA was observed in the sponge connective tissue as a result of zinc supplementation. Whereas the activity of deoxythymidine kinase was not measurable in the connective tissue during zinc-restriction phase, it became near normal following zinc supplementation.

Thus changes in the zinc concentration of plasma, erythrocytes, leucocytes, and urine and changes in the activities of zinc-dependent enzymes such as alkaline phosphatase, RNase in the plasma, and deoxythymidine kinase in the tissue during the zinc restriction phase, appear to have been induced specifically by a mild deficiency of zinc in the volunteers. One unexpected finding was with respect to plasma ammonia level which appeared to increase during the zinc-restricted period. We recently have reported a similar finding in zinc deficient rats (88). This may have important health implications in so far as zinc deficiency in man is concerned, inasmuch as in the liver disease hyperammonemia is believed to affect the central nervous system adversely.

The changes observed with respect to body weight were related to dietary zinc intake. An increased loss of fat, as determined by subscapular thickness, and normal absorption of fat during the zinc restriction phase, suggests that deficiency of zinc may have led to hypercatabolism of fat in our subjects. In experimental animals an increase in free fatty acids has been observed as a result of zinc deficiency (89). Indeed more studies are required in human subjects in order to document increased fat catabolism attributable to zinc restriction.

Direct measurements of DNA and protein synthesis in experimental animals suggest that in zinc deficiency, protein synthesis is adversely affected (90, 91). Our data in human subjects revealed that the total protein, total collagen, and RNA/DNA increased as a result of zinc supplementation. The activity of deoxythymidine kinase was not measurable during zinc-restriction phase but became 70% of normal level following zinc supplementation for three months. Similar data have been reported for experimental animals. Thus it appears that deoxythymidine kinase in human subjects is also a zinc-dependent enzyme, and an adverse effect of zinc deficiency on this enzyme may adversely affect protein synthesis. Our studies do not rule out an adverse effect of zinc deficiency on protein catabolism. Further studies are required to establish the effect of zinc restriction on protein catabolism.

These data indicate that plasma alkaline phosphatase and deoxythymidine kinase in sponge connective tissue in human subjects are zinc-dependent enzymes inasmuch as changes in their activities were related to only one dietary manipulation, namely zinc intake. Changes in the activities of plasma RNase also appear to be related to zinc intake under the conditions of our experiments. Thus the determination of the activities of these enzymes may be helpful in correlating uncomplicated zinc status in man, particularly if the changes are observed following zinc supplementation for a short period of time.

Changes in the plasma zinc concentration were observed early (within four to six weeks) and correlated with the severity of dietary zinc restriction. Thus plasma zinc may be very useful in assessment of zinc status in man, provided infections, myocardial infarction, intravascular hemolysis, and acute stress are ruled out (87). As a result of infections, myocardial infarction and acute stress, zinc from the plasma compartment may redistribute to other tissues, thus making an assessment of zinc status in the body a difficult task. Intravascular hemolysis would also spuriously increase the plasma zinc level inasmuch as the content of red-cell zinc is much higher than the plasma.

Changes in the red-cell zinc were slow to appear as expected; on the other hand, changes in the leucocyte zinc appeared more sensitive to changes in zinc intake. Urinary excretion of zinc decreased as a result of dietary zinc restriction, suggesting that renal conservation of zinc may be important for homeostatic control mechanism in man. Thus determination of zinc in 24-hour urine may be of additional help in diagnosing zinc deficiency provided cirrhosis of the liver, sickle cell disease, and chronic renal disease, are ruled out (87). These conditions are known to have hyperzincuria and associated zinc deficiency.

Our data indicate that during the zinc-deficient state the subjects were in a more positive balance for zinc. This would suggest that perhaps a test based on oral challenge of zinc and subsequent plasma zinc determination may be able to distinguish between zinc-deficient and zinc sufficient-states in human subjects.

The concentration of zinc in hair appears to reflect chronic zinc nutriture. Thus if the hair has been growing at a reasonable rate, hair zinc is a useful index of chronic zinc status in the body. Hair zinc, however, does not reflect changes in the status of zinc on an acute basis. Similar remarks apply to zinc in the red cells. Ultimately the response to therapy with zinc is probably the most reliable index for making a diagnosis of zinc-deficient state in man.

Metabolism of Zinc and Biochemistry

It is estimated that the zinc content of a normal 70-kg male is approximately 1.5–2.0 gm. Liver, kidney, bone, retina, prostate, and muscle appear to be rich in zinc. In man, zinc content of testes and skin has not been determined accurately, although clinically it appears that these tissues are sensitive to zinc depletion.

Zinc in the plasma is mostly present as bound to albumin, but other proteins, such as α_2 macroglobulin, transferrin, ceruloplasmin, haptoglobin, and gamma globulins, also bind significant amount of zinc (92). Besides protein-bound fraction, a small proportion of zinc (2–3% of total zinc) in the plasma also exists as ultrafiltrable fractions, mostly bound to amino acids, and a smaller fraction exists as ionic form. Histidine, glutamine, threonine, cystine, and lysine appear to have significant zinc-binding ability (92). Recently it has been shown that prostaglandine E_2 not only binds zinc but also facilitates its transport across the intestinal mucosa in the rat (93). Whereas amino acids competed effectively with albumin, haptoglobin, transferrin, and IgG for binding of zinc, a similar phenomenon was not observed with respect to ceruloplasmin and α_2 macroglobulin, suggesting that the latter two proteins exhibited a stronger binding affinity for zinc (92).

Approximately 20–30% of ingested dietary zinc is absorbed. Data on both the site(s) of absorption in man and the mechanism(s) of absorption, whether it be active, passive, or facultative transport, are meager. Zinc absorption is variable in extent and highly depends upon a variety of factors. Zinc is more available for absorption from animal proteins. Among other factors that might affect zinc absorption are body size, the level of zinc in the diet, and the presence in the diet of other potentially interfering substances, such as calcium, phytate, fiber, and the chelating agents.

Normal zinc intake in a well-balanced American diet with animal protein is approximately 12–15 mg/day. Urinary zinc loss is approximately 0.5 mg/day. Loss of zinc by sweat can be considerable under certain climatic conditions. Under normal conditions approximately 0.5 mg of zinc can be lost daily by sweating. Endogenous zinc loss in the gastrointestinal tract can amount to 1–2 mg/day.

Daily requirement of zinc for human subjects is not established. In view of the fact that several dietary constituents can affect the availability of zinc, it is apparent that the dietary requirement must very greatly from one region to the other, depending upon the food habits of the population.

Biochemistry. Zinc is essential for many biological functions in man and animals. Major functions of zinc in human and animal metabolism appear to be enzymatic. There are now more than 70 metalloenzymes known to require zinc for their functions (12, 13). Zinc enzymes are known to participate in a variety of metabolic processes including carbohydrate, lipid, protein, and nucleic acid synthesis or degradation. The metal is present in several dehydrogenases, aldolases, peptidases, and phosphatases.

Zinc atoms in some of the enzyme molecules participate in catalysis and also appear to be essential for maintenance of structures of apoenzymes (12, 13). For instance, alcohol dehydrogenase from horse liver contains four gram atoms of zinc per molecular weight of 80,000. Two zinc ions are essential for catalytic activity and are bound to the enzyme via two cysteinyl SH groups and the imidazole ring of a histadyl residue. The fourth metal coordination site is thought to be occupied by water. The other two zinc ions are involved in maintaining structure and each is bound to four SH groups.

A deficiency of zinc in *E. gracilis* has been shown to affect adversely all the phases of cell cycle (G_1, S, G_2, and mitosis), thus indicating that zinc is required for biochemical processes essential for cells to pass from G_2 to mitosis, from S to G_2, and from G_1 to S (13). The effect of zinc on cell cycle is undoubtedly attributable to its vital role in DNA synthesis (90, 91). Many studies have shown that zinc deficiency in animals impairs the incorporation of labeled thymidine into DNA. This effect in the animals has been detected within a few days after the zinc-deficient diet is instituted (90).

Prasad and Oberleas (90) provided evidence that decreased activity of deoxythymidine kinase may be responsible for this early reduction in DNA synthesis. As early as six days after the animals were placed on the dietary treatment, the activity of deoxythymidine kinase was reduced in rapidly regenerating connective tissue of zinc-deficient rats, compared with pair-fed controls. These results recently have been confirmed by Dreosti and Hurley (94). The activity of deoxythymidine kinase in 12-day-old fetuses taken from females exposed to a dietary zinc deficiency during pregnancy was significantly lower than in adlibitum and restricted-fed controls. Activity of the enzyme was not restored by in vitro addition of zinc, whereas addition of copper severely affected the enzyme activity adversely. Recently zinc has been shown to be an essential constituent for the DNA polymerase of E. coli (95). Whether or not this enzyme is affected adversely in animal model, because of deficiency of zinc, is not known.

Livers from zinc-deficient rats incorporated less phosphorus-32 into the nucleotides of RNA than livers from pair-fed controls, and DNA-dependent RNA polymerase has been shown to be a zinc-dependent enzyme (96, 97). The activity of RNase is increased in zinc deficient tissues (98). This suggests that the catabolism of RNA can be regulated by zinc.

From the above discussion, it appears that zinc may have its primary effect on zinc-dependent enzymes that regulate the biosynthesis and catabolic rate of RNA and DNA. In addition, zinc may also play a role in the maintenance of polynucleotide conformation. Sandstead et al. (99) observed abnormal polysome profiles in the liver of zinc-deficient rats and mice. Acute administration of zinc appeared to stimulate polysome formation both in vivo and in vitro. This finding is supported by the data of Fernandez-Madrid, Prasad, and Oberleas (42), who noted a decrease in the polyribosome content of zinc-deficient connective tissue from rats and a concomitant increase in inactive monosomes.

Enzyme Changes in Zinc Deficiency. Since zinc is required for many enzymes, it is reasonable to speculate that the level of zinc in cells controls the physiological processes through the formation and/or regulation of activity of zinc-dependent enzymes. Until 1965, there was no evidence in the literature to support this concept. During the past decade it has been shown that the activity of various zinc-dependent enzymes was reduced in the testes, bones, esophagus, and kidneys of zinc-deficient rats in comparison with their pair-fed controls (90, 91, 100, 101). These results correlated with the decreased zinc content in the above tissues of the zinc-deficient rats, thus suggesting that the likelihood of detecting any biochemical changes is greatest in tissues that are sensitive to zinc depletion.

In several studies, the activity of the alkaline phosphatase was found to be reduced in bones from zinc-deficient rats, pigs, cows, chicks, turkey poults, and quails (*91*). The activity of alkaline phosphatase can be reduced also in the intestine, kidneys, and stomach in experimental animals because of zinc deficiency. There may not only be a loss of activity attributable to a lack of sufficient zinc for maintaining the enzyme activity, but the amount of apoenzyme present appears to be diminished because of either a decreased synthesis or an increased degradation. Inasmuch as a lowering in the activity of this enzyme in intestinal tissue and in the plasma is observed before any sign of a lowered food intake is evident, it is concluded that the loss of enzyme activity is directly attributable to zinc deficiency.

Two important zinc metalloenzymes in protein digestion are the pancreatic carboxypeptidase A and B. A loss of activity of the pancreatic carboxyeptidase A in zinc deficiency is a consistent finding (*91*). According to some investigators, within two days of institution of zinc-deficient diet in rats, the enzyme lost 24% of its activity, and within three days of dietary zinc repletion, the activity of pancreatic carboxypeptidases is restored to normal levels (*91*). These results indicate that the level of food intake has no influence and that a decreased activity of carboxypeptidase A in the pancreas was related specifically to a dietary lack of zinc. As in the case of the alkaline phosphatase, the amount of carboxypeptidase A apoenzyme present appears to be diminished in zinc-deficient pancreas.

Reduced activity of carbonic anhydrase, another zinc metalloenzyme, has been reported in gastric and intestinal tissues and in erythrocytes when the activity of the enzyme was expressed per unit of erythrocytes (*91*). Recently in sickle-cell-disease patients, an example of a conditioned zinc-deficient state, the content of carbonic anhydrase in the red cells was found to be decreased, correlating with the zinc content of the red cells (*10, 75*). Inasmuch as the technique measured the apoenzyme content, it appears that zinc may have a specific effect on the synthesis of this protein by some mechanism yet to be understood.

Several investigators have now shown that zinc deficiency lowers the activity of alcohol dehydrogenase in the liver, bones, testes, kidneys, and esophagus of rats and pigs (*100, 102, 103*). In another study, alcohol dehydrogenase was assayed in subcellular fractions of liver and retina from zinc-deficient and control rats using retinol and ethanol as substrates (*103*). The activity of alcohol dehydrogenase was significantly decreased as a result of zinc deficiency in growing animals. In older rats, although no changes in liver zinc and activity of alcohol dehydrogenase were found, the retina was shown to be sensitive to the lack of zinc. These data show that zinc is required for the metabolism of vitamin A as well as the catabolism of ethanol. An attempt to demonstrate accumulation of apoenzyme of alcohol dehydrogenase in the zinc-deficient tissues was not successful.

Numerous metalloenzymes have the ability to remain functional even after the metal, which presumably is present at their active center, has been replaced by another metal (13). Thus in zinc deficiency, if the apoenzyme is synthesized, as has been observed in the case of E. coli alkaline phosphatase (13), then other metals which might have accumulated or are normally within the cell could substitute for zinc and generate an active enzyme. Although this is a possibility in the case of microorganisms, it certainly does not appear to be true in the case of experimental animals and man, in that the apoenzymes of alkaline phosphatase, carbonic anhydrase, carboxypeptidase, alcohol dehydrogenase, and deoxythymidine kinase do not accumulate in zinc-deficient tissues. Thus, one may conclude that a deficiency of zinc does specifically affect the activities of zinc-dependent enzymes in sensitive tissues.

One should not expect that zinc-dependent enzymes are affected to the same extent in all tissues of a zinc-deficient animal. Differences in the sensitivity of enzymes are evidently the result of differences in both the zinc ligand affinity of the various zinc metalloenzymes and in their turnover rates in the cells of the affected tissues. Thus it is to be expected that those zinc metalloenzymes which bind zinc with a very high affinity are still fully active even in extreme stages of zinc deficiency. The extent to which the metalloenzymes lose their activity also depends on the functional role that zinc plays in maintaining the enzyme structure. In some zinc-dependent enzymes, zinc deficiency may induce structural changes which increase the chance of degradation. The consequence is an increased turnover rate and a lower activity of the enzymes in the tissues. It has been suggested that the rapidity with which biochemical changes arise in response to zinc depletion and then disappear upon repletion helps to identify the primary site of metabolic functions of zinc. In studies applying dietary zinc depletion, the early changes in enzyme activities, which are detectable before a general depletion becomes evident from tissue zinc levels, indicate that the primary role of zinc must be associated with a tissue component of an extremely high turnover rate in that zinc is essential at a site where it is freely exchangeable.

Recently, the role of zinc in gonadal function was investigated in rats (104). The increase in luteinizing hormone (LH), follicle-stimulating hormone (FSH), and testosterone were assayed following intravenous administration of synthetic LH-releasing hormone (LH–RH), to zinc-deficient and restricted-fed control rats. Body weight gain, zinc content of testes, and their weights were significantly lower in the zinc-deficient rats as compared with the controls. The serum LH and FSH responses to LH–RH administration was higher in the zinc-deficient rats, but serum testosterone response was lower in comparison with the restricted-fed controls. These data indicate a specific effect of zinc on testes and suggest that gonadal function in the zinc-deficient state is affected through some alteration of testicular steroidogenesis.

Zinc may also intervene in nonenzymic, free-radical reactions (*105*). In particular, zinc is known to protect against iron-catalyzed, free-radical damage. It has been known that the free-radical oxidation (autoxidation) of polyunsaturated lipids is most effectively induced by the interaction of inorganic iron, oxygen, and various redox couples, and recent work suggests that this interaction underlies the pathological changes and clinical manifestations of iron toxicity. Iron-catalyzed, free-radical oxidation is known to be inhibited by zinc, ceruloplasmin, metalloenzymes (catalase, peroxidases, superoxide dismutase) and free-radical scavenging antioxidants, Vitamin E.

Carbon tetrachloride-induced liver injury is another animal model for studying free-radical injury to tissues. Animals maintained on high zinc regimen are resistant to this type of biochemical injury, thus suggesting that zinc may be protective against free-radical injury. More recent studies have shown that zinc also inhibits the analogous metromidazole-dependent, free-radical sequence.

Zinc prevents induced histamine release from mast cell (*14*). It is believed that this effect of zinc is attributable to its action on the cell membrane. Platelets are also affected by zinc ions. Collagen-induced aggregation of dog platelets and collagen or epinephrine-induced release of ^{14}C-serotonin were significantly inhibited by zinc (*14*). Supplementation of zinc in dogs effectively decreased aggragability of platelets as well as the magnitude of ^{14}C-serotonin release.

Zinc supplementation inhibits migration and other activities of macrophages and eventually of polymorphonuclear leukocytes, thus an induction of sterile inflammatory reaction; for instance, intraperitoneal injection of mineral oil to animals treated with zinc results in less cellular infiltration of peritoneal cavity by either type of inflammatory cell.

It has been speculated that zinc may form mercaptides with thiol groups of proteins, possibly linking to the phosphate moiety of phospholipids or interaction with carboxyl groups of sialic acid or proteins on plasma membranes, resulting in change of fluidity and stabilization of membranes (*14*).

There are also several enzymes attached to the plasma membrane which control the structure and function of the membrane, and the activities of these enzymes may be controlled by zinc. Adenosinetriphosphatase (ATPase) and phospholipase A_2 are known to be inhibited by zinc, and this effect may explain immobilization of energy-dependent activity of plasma membrane or increased integrity of the membrane structure (*14*).

Several receptors at the plasmatic membrane presumably function as a gate for transmitting information to intracellular space. In the case of mast cells, histamine-releasing agents seem to work through specific receptors at the membrane. Masking of such receptor site by membrane impermeable Zn:8-hydroxyquinolone would thus explain the inhibition of the release reaction.

The role of Ca^{++} in the function of cell microskeleton, represented by microtubules and microfilament, has been well documented (*14, 106, 107*). The contractile elements of this system are in some way responsible for the mobility of microorganelles and transport of granules to the membrane as well as excitability of the plasma membrane itself. Zinc may compete with calcium and thereby inhibit the calcium effect.

Zinc has been shown to improve filterability through 3.0-μm nucleopore filter of sickle cells in vitro (*106*). Improvement in filterability at low concentration of zinc suggests that the effect of zinc was on red-cell membrane. Recent studies show that the process of formation of irreversibly sickled cells involves the cell membrane. Calcium and/or hemoglobin binding may promote the formation of irreversibly sickled cells, thus hindering the filterability of such cells. Zinc may act favorably on the filterability of sickled cells by blocking the proposed calcium and/or hemoglobin binding to the membrane. The beneficial effect of zinc on sickling process has been demonstrated both in vitro and in vivo (*108*). It has been suggested recently that irreversible sickled cells are stabilized by abnormal interactions between the spectrin and/or actin components of the membrane skeleton and that disruption of these bonds by zinc may allow the skeleton to resume a normal shape (*107*).

Zinc is known to compete with cadmium, lead, copper, iron, and calcium for similar binding sites (*109*). In the future, a potential use of zinc may be to alleviate toxic effects of cadmium and lead in human subjects. Use of zinc as an antisickling agent is an example of its antagonistic effect on calcium, which is known to produce irreversible sickle cells by its action on red-cell membrane.

Therapeutic use of zinc is known to produce hypocupremia in human subjects (*110*). Whether or not zinc could be used to decrease copper load in Wilson's disease, remains to be demonstrated.

Summary

During the past two decades, essentiality of zinc for man has been established. Deficiency of zinc in man attributable to nutritional factors and several diseased states has been recognized. High phytate content of cereal proteins decreases availability of zinc, thus the prevalance of zinc deficiency is likely to be high in the population subsisting on cereal proteins mainly. Alcoholism is known to cause hyperzincuria and thus may play a role in producing zinc deficiency in man. Malabsorption, cirrhosis of the liver, chronic renal disease, and other chronically debilitating diseases may similarly induce zinc deficiency in human subjects. A severe deficiency of zinc which can be life threatening has been reported to occur in patients with acrodermatitis enteropathica, following total parenteral nutrition and penicillamine therapy. Zinc therapy to such patients is a life-saving procedure. A deficiency of zinc recently has been recognized to occur in patients with sickle cell anemia, and a beneficial effect of zinc therapy in such patients has been reported.

Growth retardation, male hypogonadism, skin changes, poor appetite mental lethargy, and delayed wound healing are some of the manifestations of chronically zinc-deficient human subjects. In severely zinc-deficient patients, dermatological manifestations, diarrhea, alopecia, mental disturbances, and intercurrent infections predominate and if untreated the condition becomes fatal. Zinc deficiency is known to affect testicular functions adversely in man and animals. This effect of zinc is at the end-organ level, and it appears that zinc is essential for spermatogenesis and testosterone steroidogenesis.

Zinc is involved in many biochemical functions. Several zinc metalloenzymes have been recognized in the past decade. Zinc is required for each step of cell cycle in microoragnisms and is essential for DNA synthesis. Thymidine kinase, DNA-dependent RNA polymerase, DNA-polymerase from various sources, and RNA-dependent DNA polymerase from viruses have been shown to be zinc-dependent enzymes. Zinc also regulates the activity of RNase, thus the catabolism of RNA appears to be zinc dependent. The effect of zinc on protein synthesis may be attributable to its vital role in nucleic acid metabolism.

The activities of many zinc-dependent enzymes have been shown to be affected adversely in zinc-deficient tissues. Three enzymes, alkaline phosphatase, carboxypeptidase, and thymidine kinase, appear to be most sensitive to zinc restriction in that their activities are affected adversely within three to six days of institution of a zinc-deficient diet to experimental animals.

Zinc atoms in some of the enzyme molecules participate in catalysis and also appear to be essential for maintenance of structure of apoenzymes. Zinc also plays a role in stabilization of biomembrane structure and polynucleotide conformation. Inasmuch as zinc appears to have a protective influence in hepatic cellular damage induced by carbon tetrachloride poisoning, it is reasonable to suggest that zinc also may have a direct effect on free radicals.

Zinc is known to compete with cadmium, lead, copper, iron, and calcium for similar binding sites. In the future, a potential use of zinc may be to alleviate toxic effects of cadmium and lead in human subjects. Use of zinc as an antisickling agent is an example of its antagonistic effect on calcium, which is known to produce irreversible sickle cells by its action on red cell membrane. Therapeutic use of zinc is known to produce hypocupremia in human subjects. Whether or not zinc could be used to decrease copper load in Wilson's disease remains to be demonstrated.

Literature Cited

1. Bertrand, G., Bhattacherjee, R. C., "L'action Combinee du Zinc et des Vitamines dans l'Alimentation des Animaux," *Acad. Sci., Paris, R.* (1934) **198**, 1823.
2. Todd, W. R., Elvehjem, C. A., Hart, E. B., "Zinc in the Nutrition of the Rat," *Am. J. Physiol.* (1934) **107**, 146.
3. Tucker, H. F., Salmon, W. D., "Parakeratosis or Zinc Deficiency Disease in Pigs, *Proc. Soc. Exp. Biol. Med.* (1955) **88**, 613.
4. O'Dell, B. L., Savage, J. E., "Potassium, Zinc and Distillers Dried Solubles as Supplement to a Purified Diet," *Poult. Sci.* (1957) **36**, 459.
5. Prasad, A. S., Halsted, J. A., Nadimi, M., "Syndrome of Iron Deficiency Anemia, Hepatosplenomegaly, Hypogonadism, Dwarfism and Geophagia," *Am. J. Med.* (1961) **31**, 532.
6. Prasad, A. S., Miale, A., Jr., Farid, Z., Sandstead, H. H., Darby, W. J., "Biochemical Studies on Dwarfism, Hypogonadism and Anemia," *Arch. Intern. Med.* (1963) **111**, 407.
7. Prasad, A. S., Miale, A., Farid, Z., Schulert, A., Sandstead, H. H., "Zinc Metabolism in Patients With the Syndrome of Iron Deficiency Anemia, Hypogonadism and Dwarfism," *J. Lab. Clin. Med.* (1963) **61**, 531.
8. Prasad, A. S., "Deficiency of Zinc in Man and Its Toxicity," *in* "Trace Elements in Human Health and Disease," A. S. Prasad, Ed., Vol. 1,, p. 1, Academic, New York, 1976.
9. Sandstead, H. H., Vo-Khactu, K. P., Solomon, N., "Conditioned Zinc Deficiencies," *in* "Trace Elements in Human Health and Disease," A. S. Prasad, Ed., Vol. 1, p. 33, Academic, New York, 1976.
10. Prasad, A. S., Abbasi, A., Ortega, J., "Zinc Deficiency in Man: Studies in Sickle Cell Disease," *in* "Zinc Metabolism: Current Aspects in Health and Disease," G. J. Brewer, A. S. Prasad, Eds., p. 211, A. R. Liss, New York, 1977.
11. Keilin, D., Mann, J., "Carbonic Anhydrase. Purification and Nature of the Enzyme," *Biochem. J.* (1940) **34**, 1163.
12. Riordan, J. F., Vallee, B. L., "Structure and Function of Zinc Metalloenzymes," *in* "Trace Elements in Human Health and Disease," A. S. Prasad, Ed., Vol. 1, p. 227, Academic, New York, 1976.
13. Riordan, J. F., "Biochemistry of Zinc," *Med. Clin. North Am.* (1976) **60**, 661.
14. Chvapil, M., "Effect of Zinc on Cells and Biomembranes," *Med. Clin. North Am.* (1976) **60**, 799.
15. Coble, Y. D., Schulert, A. R., Farid, Z., "Growth and Sexual Development of Male Subjects in an Egyptian Oasis," *Am. J. Clin. Nutr.* (1966) **18**, 421.
16. Prasad, A. S., Schulert, A. R., Miale, A., Farid, Z., Sandstead, H. H., "Zinc and Iron Deficiencies in Male Subjects with Dwarfism and Hypogonadism but Without Ancyclostomiasis, Schistosomiasis or Severe Anemia," *Am. J. Clin. Nutr.* (1963) **12**, 437.

17. Coble, Y. D., Van Reen, R., Schulert, A. R., Koshakji, R. P., Farid, Z., Davis, J. T., "Zinc Levels and Blood Enzyme Activities in Egyptian Male Subjects with Retarded Growth and Sexual Development," *Am. J. Clin. Nutr.* (1966) **19**, 415.

18. Ronaghy, H. A., Reinhold, J. G., Mahloudji, M., Ghavami, P., Fox, M. R. S., Halsted, J. A., "Zinc Supplementation of Malnourished Schoolboys in Iran: Increased Growth and Other Effects," *Am. J. Clin. Nutr.* (1974) **27**, 112.

19. Sandstead, H. H., Prasad, A. S., Schulert, A. R., Farid, Z., Miale, A., Jr., Bassily, S., Darby, W. J., "Human Zinc Deficiency, Endocrine Manifestations and Response to Treatment," *Am. J. Clin. Nutr.* (1967) **20**, 422.

20. Halsted, J. A., Ronaghy, H. A., Abadi, P., Haghshenass, M., Amirhakimi, G. H., Barakat, R. M., Reinhold, J. G., "Zinc Deficiency in Man: The Shiraz Experiment," *Am. J. Med.* (1972) **53**, 277.

21. Halsted, J. A., Smith, J. C., Jr., Irwin, M. I., "A Conspectus of Research on Zinc Requirements of Man," *J. Nutr.* (1974) **105**, 345.

22. Hambidge, K. M., Hambidge, C., Jacobs, M., Baum, J. D., "Low Levels of Zinc in Hair, Anorexia, Poor Growth, and Hypogeusia in Children," *Pediatr. Res.* (1972) **6**, 868.

23. Sandstead, H. H., "Zinc Nutrition in the United States," *Am. J. Clin. Nutr.* (1973) **26**, 1251.

24. Hambidge, K. M., Walravens, P. A., "Zinc Deficiency in Infants and Preadolescent Children," *in* "Trace Elements in Human Health and Disease," A. S. Prasad, Ed., Vol. 1, p. 21, Academic, New York, 1976.

25. Strain, W. H., Steadman, L. T., Lankau, C. A., Jr., Berliner, W. P., Pories, W. J., "Analysis of Zinc Levels in Hair for the Diagnosis of Zinc Deficiency in Man," *J. Lab. Clin. Med.* (1966) **68**, 244.

26. Strain, W. H., Lascari, A., Pories, W. J., Zinc Deficiency in Babies," *Proc. Int. Congr. Nutr., 7th, 1966,* 749.

27. Berfenstam, R., "Studies on Blood Zinc," *Acta Paediatr., Stockholm* (1952) **41**, 5.

28. Hellwege, H. H., "Der serumzinkspiegel and seine Veranderungen bei einigen Krankheiten im Kindesalter," *Monatsschr. Kinderheilkd.* (1971) **119**, 37.

29. Walravens, P. A., Hambidge, K. M., "Nutritional Zinc Deficiency in Infants and Children," *in* "Zinc Metabolism, Current Aspects in Health and Disease," G. J. Brewer, A. S. Prasad, Eds., p. 61, A. R. Liss, New York, 1977.

30. Sullivan, J. F., "Effect of Alcohol on Urinary Zinc Excretion," *Q. J. Stud. Alcohol* (1962) **23**, 216.

31. Gudbjarnason, S., Prasad, A. S., "Cardiac Metabolism in Experimental Alcoholism," *in* "Biochemical and Clinical Aspects of Alcohol Metabolism," V. M. Sardesai, Ed., p. 266, Charles C. Thomas, Springfield, IL, 1969.

32. Allan, J. G., Fell, G. S., Russel, R. I., "Urinary Zinc in Hepatic Cirrhosis," *Scott. Med. J.* (1975) **109**.

33. Weismann, K., Roed-Petersen, J., Hjorth, N., Kopp, H., "Chronic Zinc Deficiency Syndrome in a Beer Drinker with a Billroth II Resection," *Int. J. Dermatol.* (1976) **15**, 757.

34. Vallee, B. L., Wacker, W. E. C., Bartholomay, A. F., Robin, E. D., "Zinc Metabolism in Hepatic Dysfunction. I. Serum Zinc Concentrations in Laennec's Cirrhosis and Their Validation by Sequential Analysis," *N. Engl. J. Med.* (1956) **255**, 403.

35. Morrison, S. A., Russell, R. M., Carney, E. A., Oaks, E. V., "Zinc Deficiency: A Cause of Abnormal Dark Adaptation in Cirrhotics," *Am. J. Clin. Nutr.* (1978) **31**, 276.

36. Saldeen, T., Brunk, U., "Enzyme Histochemical Investigations on the Inhibitory Effect of Zinc on the Injurious Action of Carbon Tetrachloride on the Liver," *Frankf. Z. Pathol.* (1967) **76**, 419.

37. Mansouri, K., Halsted, J., Gombos, E. A., "Zinc, Copper, Magnesium, and Calcium in Dialysed and Non-dialysed Uremic Patients," *Arch. Intern. Med.* (1970) **125**, 88.

38. Halsted, J. A., Smith, J. C., Jr., "Plasma-zinc in Health and Disease," *Lancet* (1970) **1**, 322.

39. Husain, S. L., "Oral Zinc Sulfate in Leg Ulcers," *Lancet* (1969) **1**, 1069.

40. Serjeant, G. R., Galloway, R. E., Gueri, M. C., "Oral Zinc and Sulphate in Sickle-cell Ulcers," *Lancet* (1970) **2**, 891.

41. Greaves, M. W., "Zinc in Cutaneous Ulceration Due to Vascular Insufficiency," *Am. Heart J.* (1972) **83**, 716.

42. Fernandez-Madrid, F., Prasad, A. S., Oberleas, D., "Effect of Zinc Deficiency on Nucleic Acids, Collagen, and Noncollagenous Protein of the Connective Tissue," *J. Lab. Clin. Med.* (1973) **82**, 951.

43. Greaves, M. W., Boyde, T. R. C., "Plasma Zinc Concentrations in Patients with Psoriasis, other Dermatosis, and Venous Ulcerations," *Lancet* (1967) **2**, 1019.

44. Greaves, M. W., "Zinc and Copper in Psoriasis," *Br. J. Dermatol.* (1972) **86**, 439.

45. Portnoy, B., Molokhia, M. M., "Zinc and Copper in Psoriasis," *Br. J. Dermatol.* (1971) **85**, 497.

46. Portnoy, B., Molokhia, M. M., "Zinc and Copper in Psoriasis," *Br. J. Dermatol.* (1972) **86**, 205.

47. Pories, W. J., Henzel, J. H., Rob, C. G., Strain, W. H., "Acceleration of Wound Healing in Man with Zinc Sulfate Given by Mouth," *Lancet* (1967) **1**, 121.

48. Pories, W. J., Strain, W. H., "Zinc and Wound Healing," *in* "Zinc Metabolism," A. S. Prasad, Ed., p. 378, Charles C. Thomas, Springfield, IL, 1966.

49. Cohen, C., "Zinc Sulfate and Bedsores," *Br. Med. J.* (1968) **2**, 561.

50. Greaves, M. W., Skillen, A. W., "Effects of Long-continued Ingestion of Zinc Sulphate in Patients with Venous Leg Ulceration," *Lancet* (1970) **2**, 889.

51. Barcia, P. J., "Lack of Acceleration of Healing with Zinc Sulfate," *Ann. Surg.* (1970) **172**, 1048.

52. Myers, M. B., Cherry, G., "Zinc and the Healing of Chronic Leg Ulcers," *Am. J. Surg.* (1970) **120**, 77.

53. Clayton, R. J., "Double-blind Trial of Oral Zinc Sulphate in Patients with Leg Ulcers," *Br. J. Clin. Pract.* (1972) **26**, 368.

54. Greaves, M. W., Ive, F. A., "Double-blind Trial of Zinc Sulphate in the Treatment of Chronic Venous Leg Ulceration," *Br. J. Dermatol.* (1972) **87**, 632.

55. Hallbook, T., Lanner, E., "Serum Zinc and Healing of Venous Leg Ulcers," *Lancet* (1972) **2**, 780.

56. Oberleas, D., Seymour, J. K., Lenaghan, R., Hovanesian, J., Wilson, R. F., Prasad, A. S., "Effect of Zinc Deficiency on Wound Healing in Rats," *Am. J. Surg.* (1971) **121**, 566.

57. Elias, S., Chvapil, M., "Zinc and Wound Healing in Normal and Chronically Ill Rats," *J. Surg. Res.* (1973) **15**, 59.

58. Greene, H. L., "Trace Metals in Parenteral Nutrition," *in* "Zinc Metabolism, Current Aspects in Health and Disease," G. J. Brewer, A. S. Prasad, Eds., p. 87, A. R. Liss, New York, 1977.

59. Klingberg, W. G., Prasad, A. S., Oberleas, D., "Zinc Deficiency Following Penicillamine Therapy," *in* "Trace Elements in Human Health and Disease," Vol. 1, p. 51, A. S. Prasad, Ed., Academic, New York, 1976.

60. Pidduck, H. G., Wren, P. J. J., Price Evans, D. A., "Plasma Zinc and Copper in Diabetes Mellitus," *Diabetes* (1970) **19**, 234.

61. Henzel, J. H., DeWeese, M. S., Lichti, E. L., "Zinc Concentrations within Healing Wounds," *Arch. Surg., Chicago* (1970) **100**, 349.

62. Cuthbertson, D. P., Fell, G. S., Smith, C. M., Tolstone, W. J., "Metabolism after Injury. I. Effects of Severity, Nutrition, and Environmental Temperature on Protein, Potassium, Zinc, and Creatine," *Br. J. Surg.* (1972) **59**, 925.

63. Pekarek, R. S., Wannemacher, R. W., Beisel, W. R., "The Effect of Leukocyte Endogenous Mediator (LEM) on the Tissue Distribution of Zinc and Iron," *Proc. Soc. Exp. Biol. Med.* (1972) **140**, 685.

64. Simkin, P. A., "Zinc Sulphate in Rheumatoid Arthritis," *in* "Zinc Metabolism, Current Aspects in Health and Disease," G. J. Brewer, A. S. Prasad, Eds., p. 343, A. R. Liss, New York, 1977.

65. Henkin, R. I., Marshall, J. R., Meret, S., "Maternal-fetal Metabolism of Copper and Zinc at Term," *Am. J. Obstet. Gynecol.* (1971) **110**, 131.

66. Sandstead, H. H., Glasser, S. R., Gillespie, D. D., "Zinc Deficiency: Effect on Fetal Growth, Zinc Concentration and ^{65}Zinc Uptake," *Fed. Proc. Fed. Am. Soc. Exp. Biol.* (1970) **29**, 297.

67. Hurley, L. S., "Perinatal Effects of Trace Element Deficiencies," *in* "Trace Elements in Human Health and Disease," Vol. 2, p. 301, A. S. Prasad, Ed., Academic, New York, 1976.

68. Caldwell, D. F., Oberleas, D., Clancy, J. J., Prasad, A. S., "Behavioral Impairment in Adult Rats Following Acute Zinc Deficiency," *Proc. Soc. Exp. Biol. Med.* (1970) **133**, 1417.

69. Halas, E. S., Rowe, M. C., Orris, R. J., McKenzie, J. M., Sandstead, H. H., "Effects of Intra-uterine Zinc Deficiency on Subsequent Behavior," *in* "Trace Elements in Human Health and Disease," A. S. Prasad, Ed., Vol. I, p. 327, Academic, New York, 1976.

70. Lökken, P. M., Halas, E. S., Sandstead, H. H., "Influence of Zinc Deficiency on Behavior," *Proc. Soc. Exp. Biol. Med.* (1973) **144**, 680.

71. Hurley, L. S., Shrader, R. E., "Congenital Malformations of the Nervous System of Zinc Deficient Rats," *Int. Rev. Neurobiol., Suppl.* (1972) **1**, 7.

72. Damyanov, I., Dutz, W., "Anencephaly in Shiraz, Iran," *Lancet* (1971) **1**, 82.

73. Halsted, J. A., Hackley, B. M., Smith, J. C., Jr., "Plasma-zinc and Copper in Pregnancy and after Oral Contraceptives," *Lancet* (1968) **2**, 278.

74. Prasad, A. S., Obrleas, D., Lei, K. Y., Moghissi, K. S., Stryker, J. C., "Effect of Oral Contraceptive Agents on Nutrients. I. Minerals," *Am. J. Clin. Nutr.* (1975) **28**, 377.

75. Prasad, A. S., Schoomaker, E. B., Ortega, J., Brewer, G. J., Oberleas, D., Oelshlegel, F. J., "Zinc Deficiency in Sickle Cell Disease," *Clin. Chem., Winston-Salem, NC* (1975) **21**, 582.

76. Oelshlegel, F. J., Brewer, G. J., Knutsen, C., Prasad, A. S., Schoomaker, E. B., "Studies on the Interaction of Zinc with Human Hemoglobin," *Arch. Biochem. Biophys.* (1974) **163**, 742.

77. Oelshlegel, F. J., Brewer, G. J., Prasad, A. S., Knutsen, D., Schoomaker, E. B., "Effect of Zinc on Increasing Oxygen Affinity of Sickle and Normal Red Blood Cells," *Biochem. Biophys., Res. Commun.* (1973) **53**, 560.

78. Brewer, G. J., Oelshlegel, F. J., Jr., Prasad, A. S., "Zinc in Sickle Cell Anemia," *in* "Erythrocyte Structure and Function," G. J. Brewer, Ed., Vol. 1, p. 417, Alan R. Liss, New York, 1975.

79. Danbolt, N., Closs, K., "Akrodermatitis Enteropathica," *Acta Derm. Venereol.* (1942) **23**, 127.
80. Perry, H. O., "Acrodermatitis Enteropathica," *in* "Clinical Dermatology," D. J. Demis, R. G. Crounse, R. L. Dobson, J. McGuire, Eds., Vol. 1, p. 1, Harper and Row, New York, 1974.
81. Dillaha, C. J., Lorincz, A. L., Aavick, O. R., "Acrodermatitis Enteropathica," *JAMA* (1953) **152**, 509.
82. Moynahan, E. J., "Acrodermatitis Enteropathica with Secondary Lactose Intolerance and Tertiary Deficiency State, Probably Due to Chelation Essential Nutrients by Diiodohydroxyquinolone," *Proc. R. Soc. Med.* (1966) **59**, 7.
83. Moynahan, E. J., Barnes, P. M., "Zinc Deficiency and a Synthetic Diet for Lactose Intolerance," *Lancet* (1973) **1**, 676.
84. Barnes, P. M., Moynahan, E. J., "Zinc Deficiency in Acrodermatitis Enteropathica: Multiple Dietary Intolerance Treated with Synthetic Diet," *Proc. R. Soc. Med.* (1973) **66**, 327.
85. Prasad, A. S., "Metabolism of Zinc and Its Deficiency in Human Subjects," *in* "Zinc Metabolism," A. S. Prasad, Ed., p. 250, Charles C. Thomas, Springfield, IL, 1966.
86. Brummerstedt, E., Flagstad, T., Andresen, E., "The Effect of Zinc on Calves with Hereditary Thymus Hypoplasis (Lethal Tract A 46)," *Acta Pathol. Microbiol. Scand. Sect. A* (1971) **79**, 686.
87. Prasad, A. S., Rabbani, P., Abbasi, A., Bowersox, E., Fox, M. R. S., "Experimental Zinc Deficiency in Man," *Ann. Intern. Med.*, in press.
88. Rabbani, P., Prasad, A. S., "Blood Urea Nitrogen (BUN), Plasma Ammonia and Liver Ornithine-trans-carbamylase Activity in Zinc Deficient Rats," *Am. J. Physiol.* (1978) **235**: E 203–206.
89. Underwood, E. J., "Zinc," *in* "Trace Elements in Human and Animal Nutrition," E. J. Underwood, Ed., p. 196, Academic, New York, 1977.
90. Prasad, A. S., Oberleas, D., "Thymidine Kinase Activity and Incorporation of Thymidine into DNA in Zinc-deficient Tissue," *J. Lab. Clin. Med.* (1974) **83**, 634.
91. Kirchgessner, M., Roth, H. P., Weigand, E., "Biochemical Changes in Zinc Deficiency," *in* "Trace Elements in Human Health and Disease," A. S. Prasad, Ed., Vol. 1, p. 189, Academic, New York, 1976.
92. Prasad, A. S., Oberleas, D., "Binding of Zinc to Amino Acids and Serum Proteins in Vitro," *J. Lab. Clin. Med.* (1970) **76**, 416.
93. Song, M. K., Adham, N. F., "Role of Prostaglandin E_2 in Zinc Absorption in the Rat," *Am. J. Physiol.* (1978) **234**, E99.
94. Dreosti, I. E., Hurley, L. S., "Depressed Thymidine Kinase Activity in Zinc-deficient Rat Embryos," *Proc. Soc. Exp. Biol. Med.* (1975) **150**, 161.
95. Slater, J. P., Mildvan, A. S., Loeb, L. A., "Zinc in DNA Polymerase," *Biochem. Biophys. Res. Commun.* (1971) **44**, 37.
96. Terhune, M. W., Sandstead, H. H., "Decreased RNA Polymerase Activity in Mammalian Zinc Deficiency," *Science* (1972) **177**, 68.
97. Scrutton, M. C., Wu, C. W., Goldthwait, D. A., "The Presence and Possible Role of Zinc in RNA Polymerase Obtained from *Escherichia coli,*" *Proc. Nat. Acad. Sci. U.S.A.* (1971) **68**, 2497.
98. Prasad, A. S., Oberleas, D., "Ribonuclease and Deoxyribonuclease Activities in Zinc-deficient Tissues," *J. Lab. Clin. Med.* (1973) **82**, 461.
99. Sandstead, H. H., Hollaway, W. L., Baum, V., "Zinc Deficiency: Effect on Polysomes," *Fed. Proc. Fed. Am. Soc. Exp. Biol.* (1971) **30**, 517.
100. Prasad, A. S., Oberleas, D., Wolf, P., Horwitz, J. P., "Studies on Zinc Deficiency: Changes in Trace Elementsand Enzyme Activities in Tissues of Zinc-deficient Rats," *J. Clin. Invest.* (1967) **46**, 549.

101. Prasad, A. S., Oberleas, D., Miller, E. R., Luecke, R. W., "Biochemical Effects of Zinc Deficiency: Changes in Activities of Zinc-dependent Enzymes and Ribonucleic Acid and Deoxyribonucleic Acid Content of Tissues," *J. Lab. Clin. Med.* (1971) **77**, 144.
102. Prasad, A. S., Oberleas, D., "Changes in Activity of Zinc-dependent Enzymes in Zinc-deficient Tissues of Rats," *J. Appl. Physiol.* (1971) **31**, 842.
103. Huber, A. M., Gershoff, S. N., "Effects of Zinc Deficiency on the Oxidation of Retinol and Ethanol in Rats," *J. Nutr.* (1975) **105**, 1486.
104. Lei, K. Y., Abbasi, A., Prasad, A. S., "Function of Pituitary-gonadal Axis in Zinc-deficient Rats," *Am. J. Physiol.* (1976) **230**, 1730.
105. Editorial, "A Radical Approach to Zinc," *Lancet* (1978) **1**, 191.
106. Dash, S., Brewer, G. J., Oelshlegel, F. J., Jr., "Effect of Zinc on Haemoglobin Binding by Red Blood Cell Membranes," *Nature* (1974) **250**, 251.
107. Lux, S. E., John, K. M., "Unsickling of 'Irreversibly' Sickled Ghosts by Conditions which Interfere with Spectrin-actin Polymerization," *Pediatric Res.* (1978) **12**, abstract no. 630.
108. Brewer, G. J., Brewer, L. F., Prasad, A. S., "Suppression of Irreversibly Sickled Erythrocytes by Zinc Therapy in Sickle Cell Anemia," *J. Lab. Clin. Med.* (1977) **90**, 549.
109. Hill, C. H., "Mineral Interrelationships," *in* "Trace Elements in Human Health and Disease," A. S. Prasad, Ed., Vol. II, p. 281, Academic, New York, 1976.
110. Prasad, A. S., Brewer, G. J., Schoomaker, E. B., Rabbani, P., "Hypocupremia Induced by Large Doses of Zinc Therapy in Adults," *JAMA*, in press.

RECEIVED July 10, 1978. In part supported by a grant from NIH, NIAMD (AM-19338), a contract from FDA, a comprehensive sickle cell center grant (USPHS-HL-16008) to Wayne State University and a grant from Meyer Laboratories, Ft. Lauderdale, Florida.

14

Serum Copper in Relation to Age

ANIECE A. YUNICE

Medical Service, Oklahoma City, Veterans Administration Hospital, and
Department of Medicine, University of Oklahoma, Health Sciences Center,
Oklahoma City, OK 73104

*An investigation was undertaken to determine serum copper
and ceruloplasmin concentrations in 180 normal male and
44 female subjects ranging in age between 20 and 89 years.
In the female population, correlation of serum copper with
age was not statistically significant whereas a marginal but
significant increase was observed in the male population.
In neither of the two groups was there any significant alteration of ceruloplasmin with age. While serum ceruloplasmin
level was significantly higher in the females, no such significant difference could be observed in serum copper levels
between males and females. The significance of copper in
the environment with relation to other trace and ultratrace
metals and the role they play in the aging processes is
reviewed in the light of current literature.*

It is only recently that ultratrace metals have assumed a vital role in the
biological sciences in such areas as aging and modern chronic diseases.
Two factors have contributed to the advancement of our knowledge in
this important discipline: the advent of new technology and the altered
environment of soil, water, and air. The new technology has provided
us with tools to deal with the problem of detection and quantification at
the picogram level of most elements in various biological materials and
hence has enabled us to better define the biochemical role of hitherto
biologically unknown ultratrace metals. The altered environment, on the
other hand, has created an ecological imbalance of distribution of natu-
rally existing metals in our food, water, and air. Very little is known
about the influence of these metal contaminants on the aging processes.
This review will attempt to examine the role of the essential element,

0-8412-0416-0/79/33-172-230$07.25/0

copper, in the mammalian organism within the scope of other essential ultratrace metals and the homeostatic mechanism that controls its metabolism in the aging process with particular reference to our finding of a significant correlation between serum copper and aging (*1*). Readers interested in more information on copper metabolism and its homeostatic regulation in biological systems are referred to the following excellent articles (*2–16*).

Role of Ultratrace Metals in the Biological System

Essential and Nonessential Metals. It is well known that elements in the biological systems may vary a great deal in their concentration from organ to organ and from species to species, but for the purpose of this chapter, the following classification of elemental concentrations has been adopted (*17*): major, $\geq 1\%$; minor, $0.10–1\%$; micro, $0.01–0.1\%$; trace, $0.01–0.001\%$; ultratrace, $< 0.001\%$. Since total copper in the average, "standard man" (*18*) is approximately 150 mg (*2*), its classification would fall between trace and ultratrace concentration. However, as is the case with any other element, what is a trace in one organ may be an ultratrace in another, but for serum copper concentration, which is about 100 μg%, the definition of copper as an ultratrace metal by the above classification may not be justifiable. If the criteria of the "standard man" is taken into account, however, the definition seems appropriate.

The bioconcentration of ultratrace metals with age is largely unexplored, particularly metals of known biological function. It is generally accepted by most workers in this field that metals can be classified into essential, nonessential, and inert. In the absence of a genetic disorder, the bioconcentrations of essential ultratrace metals such as zinc, copper, iron, manganese, cobalt, and molybdenum are normally maintained within a narrow range by homeostatic mechanism in affecting intestinal absorption and biliary and urinary excretion. Age-connected disorders other than those genetically invoked, however, can sometimes disrupt the homeostatic mechanism. Large dietary intake, dysfunction of the gastrointestinal tract, and renal insufficiency, for example, can contribute to metal accumulation above the physiological range. No similar mechanism of homeostatic control appears to be operative to prevent accumulation of environmental contaminants for such toxic metals such as mercury, cadmium, and lead.

Copper in the Environment. Copper was probably the first metal to be used by man (*15*). Its use in the electric industry has been accelerated enormously in the last century because of its good electric conductivity. Fifty percent of the copper produced is used for electrical purposes and 15% is used for building construction, including pipes, plumbing, and roofing. Copper is also incorporated in various alloys such as brass, which is a combination of copper and zinc, and bronze, which is a copper–tin alloy.

Human exposure to excessive copper intake from a variety of sources has been on the increase in recent years. Most of the sources are attributable to modern technology and lifestyle. Many examples could be cited in food and water contamination: increased use of water pipes made of copper (15) in most municipal water systems (19); dispensing machines with copper check valves (20); acidic food or drink in metal containers (21); and semipermeable membranes containing copper or copper tubing used in hemodialysis units (22, 23).

As for the sources of contamination from air which are likely to cause respiratory ailments, the primary contributor is fungicide sprays containing copper sulfate solutions (24). Another major source is copper mines and copper industries, where copper concentration in the environment may reach as high as 1% (13). A most recent contributor to serious copper exposure in the female is the copper-containing intrauterine contraceptive device (25, 26, 27).

Although iron was the first metal to be recognized for its essentiality for the mammalian organism, copper was chronologically the second in line to be recognized as such. Its ubiquity in all living organisms, in food and in the environment, exceeds any other known essential ultratrace metal. Early in the nineteenth century its presence in a number of animals, particularly marine organisms, was clearly demonstrated. For example, hemocyanin, the blue respiratory enzyme, was recognized to be a copper enzyme present in the *Mollusca* (28). Essentiality of copper in the rat was demonstrated by Hart (29) and subsequently other workers documented it in other species as well (30, 31).

Role of Copper in the Enzyme System. It is now well recognized among workers in the ultratrace-metal field that copper is an essential nutrient for all forms of life, being the vital constituent of all living cells (32). The essentiality of copper is by virtue of its ability to catalyze biological oxidation whether the copper is in the protein-bound or ionic form, although it is likely that the ionic copper is able to do this more efficiently than the protein-bound copper (33).

Copper plays an important role in the constituents of many enzymes of the mammalian organism as well as in plants and anthropods. Several classes of oxidizing enzymes for copper have been described, including the cytochrome oxidases which are the terminal oxidases in the mitochondrial electron transport system, a key reaction in energy metabolism (34), and the amine oxidases (35) of which there are a number that contain copper (36, 37, 38). Lysyl oxidase (39) is probably the most important since it plays a major role in elastin and collagen synthesis

(*30*). Dopamine-β-hydroxylase is another copper enzyme that plays a major function in the biosynthesis of norepinephrine (*40*). A protective catalyst, superoxide dismutase, has been described to catalyze the dismutation of the superoxide anion free radical (*41*). Other enzymes known to contain copper are the laccases, the phenol oxidases, and the ascorbic acid oxidases (*12*).

Several forms of copper proteins have been reported to exist, the most important of which is ceruloplasmin. In addition, there are known to be specific copper-binding proteins such as erythrocuprein (in red blood cells), cerebrocuprein (in brain cells), and hepatocuprein (in liver cells (*42, 43, 44*).

These proteins have recently been found to be identical to superoxide dismutase, an enzyme that catalyzes the dismutation of the superoxide anion free radical (*41, 45*). However, since its discovery is of recent undertaking, its deficiency in the mammalian organism has not been well defined. It is believed to play an important catalytic role in the decomposition of two anions (O_2^-) and two H^+ to form O_2 and H_2O_2, thus protecing the aerobic organisms from the deleterious effect of free radicals. They all contain copper and zinc. A manganese-containing superoxide dismutase has recently been isolated from prokaryocytes (*46*).

Ceruloplasmin, first reported by Holmberg and Laurrell (*47*), is a copper-alpha-2-globulin. It comprises about 95% of the total copper body pool and is released only when the protein molecule is catabolized. This fraction of copper is in an exchangeable equilibrium with the ionic form; the remainder is loosely bound to albumin (*48*) and to amino acids (*49*). This last fraction recently has been reported to play a major role in transporting copper in the body.

Functional Changes in the Copper Enzyme Components

Diseases of Copper Deficiency. Manifestations of copper deficiency have been described by several authors (*16, 32*). The most common features in animals and man are anemia, ataxia, cardiovascular rupture, skeletal deformities, and achromotricia. Three systems in the body seem to suffer the consequences of copper deficiency: the hemopoietic system, the vascular and skeletal system, and the central nervous system.

It has been known for some time that copper deficiency leads to anemia and failure of the erythropoietic system to mature (*50, 51*). Although the exact mechanism involved is still not well defined, recent evidence suggests that copper may be essential for iron absorption and mobilization for hemoglobin synthesis. A ferrous-to-ferric cycle with respect to the role of copper in iron metabolism has been proposed by several workers. Role of ceruloplasmin in the spontaneous oxidation of ferrous to ferric was first suggested by Curzon and O'Reilly (52). This was subsequently confirmed by Osaki et al. (53) who also proposed the term to be renamed ferroxidase.

According to this theory, which has been recently decribed by Frieden (54), the rate-limiting step in the events that lead to hemoglobin synthesis is the spontaneous oxidation of ferrous (Fe^{2+}) to the ferric (Fe^{3+}) state in the plasma where it is transported with transferrin (55). The oxidative step highly depends on the copper-containing enzyme, ferroxidase. Although ceruloplasmin possesses ferroxidase activity, it is unlikely that it can perform synthetic function in the hemoglobin molecule. There are, however, certain difficulties in reconciling this theory with certain established facts (10). The first is lack of evidence for the presence of a defective iron metabolism in Wilson's Disease in spite of low ceruloplasmin level. An alternate iron pathway which is independent of ceruloplasmin has been suggested to be present in these subjects. The only abnormality of iron metabolism described in Wilson's Disease to date is iron deficiency (56). The other fact is that rat ceruloplasmin was found to be devoid of all ferroxidase activity or very low activity if any at all (10). These two facts taken together would throw some doubt on the validity of this theory.

An alternate hypothesis has been proposed recently by Lee et al. (10). It postulates that perhaps ceruloplasmin competes for specific binding sites occupied by ferrous iron on the reticuloendothelial cell surface, thus forming an iron–ceruloplasmin compound which in turn acts as an intermediary in transferring iron to transferrin. The validity of this hypothesis is being tested by the authors at this time.

Hepatic mitochondria isolated from copper-deficient animals were found to be deficient in the cytochrome oxidase activity which correlated well with hem synthesis (57). Failure to synthesize hem from ferric iron and protoporphyrin could be enhanced by succinate or inhibited by cyanide, which suggests that the reduction from ferric to ferrous requires an intact electron transport system in order for hem synthesis to go into completion.

Major abnormalities in vascular and skeletal systems have been originally observed in animals, leading to the recognition of the role of copper in maintaining the integrity of connective-tissue metabolism and the myelination of the central nervous system. Cardiac tissue abnormality was first described in cattle and was called "falling disease." The most prominent features of the disease are the fibrotic tissue of the cardiac muscle, atrophy of the heart, and dissecting aneurism of the aorta (58). Subsequent work led to similar findings in the rat, pig, and chicken (31, 59). Collagen cross-linking recently has been implicated as the mechanism responsible for the weakened tensile strength of the vascular system in copper-deficient animals. Lysyl oxidase, as mentioned previously, is a copper-dependent enzyme which is responsible for maintaining the cross-linking characteristics of the collagen molecule (60).

Failure of the neuron to myelinate because of copper deficiency leads to abnormalities of the nervous system. This was first described in lambs and was referred to as "swaybacks." It is not known whether this is attributable to a diminished level of tyrosine hydroxylase, which is a copper-dependent enzyme. It is known, however, that epinephrine and norepinephrine are decreased in animals that are deficient in copper (11). In ruminants, copper is shown to be necessary for myelination of nerves and for the maintenance of normal skin pigmentation (33, 61).

Except in genetic disorders such as Menkes Syndrome, caused by abnormalities in protein metabolism in children (62), hypocupremia has not been demonstrated to occur in man (49). Two other disorders, nephrosis and iron deficiency, are known to have low serum copper concentrations.

Copper deficiency in human subjects recently has been reviewed by Graham and Cordano (9). Previously it has been widely accepted that copper deficiency in man does not occur (63), but recent evidence indicates that its presence is prevalent in untreated malnourished infants (64) and in infants fed low copper-milk diet (65, 66). In addition, copper deficiency has been described in premature infants (67), in malnourished children during hyperalimentation (68, 69), and in adults (70, 71, 72).

The genetic syndrome referred to as Menkes kinky hair has been studied by Williams et al. (73). No changes were observed following oral copper supplementation in plasma copper concentration or ceruloplasmin concentration in these patients. However, when copper was given intravenously, a rise in ceruloplasmin was seen. The results indicate that in Menkes syndrome, there is a defect in copper-binding protein which leads to excessive losses of copper from the gastrointestinal tract.

Studies in these deficiency states have shed some light on the role of copper in erythropoiesis and its relationship to iron in the synthetic process. They seem to indicate that copper-containing ferroxidase is required for iron absorption and mobilization in the ferrous-to-ferric cycle oxidation. Response to copper supplementation seems to depend a great deal on whether copper deficiency is coexistent with iron deficiency or whether the deficiency has developed after iron deficiency, as evidenced by lack of erythropoietic maturation. Favorable response usually occurs in the former but not in the latter (9).

Disorders of Copper Accumulation. One of the most enigmatic diseases of copper accumulation is Wilson's Disease where copper accumulates in the liver, brain, eyes, and kidneys, leading to symptoms of hepatic cirrhosis, demyelination, and brain damage (74). This results from decreased serum ceruloplasmin levels, although the total blood copper in these patients is markedly reduced (75).

It was first suggested by Uzman et al. (76) that the defect in Wilson's Disease was primarily attributable to the synthesis of an abnormal protein with affinity for copper. Over the years, our understanding of the disease has improved, and criteria designed to establish diagnostic tools were thought to be specific. For example, it has been widely accepted that the primary characteristics of the disease are deficiency of ceruloplasmin (77), excess hepatic copper (78, 79, 80, 81), and the presence of Kayser–Fleischer rings (82). However, recent evidence suggests that these tests are not specific for Wilson's Disease in view of the fact that these abnormalities can be seen in various other dysfunctions. For example, pigmented corneal rings have been found in primary biliary cirrhosis (83), excessive copper accumulation in the liver has been documented in chronic active liver disease (84), urinary copper excretion and hepatic copper concentration were found to be elevated in primary biliary cirrhosis, and disorders of the biliary tract (85), and manifestations of hepatic disorders of Wilson's Disease have been seen in chronic active hepatitis (86).

Quite recently, a more specific test has been reported to distinguish between Wilson's Disease and primary biliary cirrhosis (87). It is based on the magnitude of radioactive copper incorporation in the liver. An impairment of the rapid pathway of copper incorporation into ceruloplasmin in subjects with primary biliary cirrhosis is the major distinguishing feature from that of Wilson's Disease. Previously, it was shown that uptake of copper by the liver in Wilson's Disease is significantly greater than uptake in normal subjects and that in heterozygotes, uptake is intermediate between the two (88).

It has been known for some time that Wilson's Disease in inherited as an autosomal recessive trait and that its prevalence in the general population amounts to one in 200,000 (82, 89, 90). The cause for the ceruloplasmin deficiency seems to be attributable to impairment of the lysosomal ability to excrete copper into bile, which is the major excretory route for copper (80, 81). Bile pigments are known to be good copper chelators (91). In addition, a copper metallothionein with high binding capacity for copper has been identified in livers of subjects with Wilson's Disease (92) and in human fetal liver (93).

It is believed that as the disease progresses, binding affinity of copper protein is diminished, thus resulting in the release of large amounts of copper into plasma, causing severe hemolysis. Some of the excess copper finds its way into the vascular system where it is deposited in such vital organs as kidney, brain, corneas, and liver.

Efforts have been made to design an experimental animal model for Wilson's Disease, but to date attempts have not been successful. Recently, however, Owen and Hazelrig (94) loaded rats with radioactive copper and measured biliary release of both copper and ceruloplasmin and compared it with hepatic and urinary copper concentration. Biliary route was reduced at the same time that hepatic and urinary copper was increased.

In man, serum copper concentrations are reported to be increased in a number of chronic illnesses such as coronary atherosclerosis with or without infarction, cerebral atherosclerosis, essential hypertension, diabetes mellitus, chronic pulmonary diseases, and various hematological disorders (95, 96). Diseases of hypercupremia and their manifestation have been described by Adelstein and Vallee (2).

Hormonal Control of Copper Metabolism.   Recent evidence suggests that metabolism of copper as well as zinc is controlled by the pituitary–adrenal axis (97, 98, 99). For example, administration of large doses of corticosteroids in humans leads to rapid and sustained drop in serum copper and zinc. Pregnant women and women on oral contraceptive agents show a significant increase in serum copper levels (100, 101, 102).

Thyroid, adrenal, and pituitary hormones are known to influence copper metabolism indirectly, presumably by reducing its biliary excretion, whereas hormones of the gonads, particularly estrogen, increase ceruloplasmin by synthesizing the protein in a manner that is independent of copper concentration in the liver (103). Our own observations on the relationship of hormones and ultratrace-metals metabolism will be mentioned later in this review. Suffice it to say that aging and its influence on mineral metabolism in general may be mediated through altered hormonal activity.

Mineral Metabolism and the Aging Process

Minerals in General.   The clinical and experimental manifestations of acute deficiencies and overt toxicities now have been well defined for most of the ultratrace metals in man and animals, but little has been done to delineate the influence of variations of exposure, absorption, and excretion of metals over the lifetime of the individual on the aging processes. It might well be that excessive loss or accumulation by long-time exposure to certain metals might interfere with the general well-being of the aging patient, might accelerate the aging process, or might affect adversely the progression of certain chronic degenerative disorders.

Certain defined criteria have been suggested as necessary for any metal to meet the requirements needed to make it a likely candidate for age-linked accumulation in man (104). First, it has to be relatively toxic if given in large amounts; second, its presence in the environment must be widely spread; third, chemically, the metal has to be highly reactive; fourth, it must have the capability of competing for the same binding sites with other metals.

Very little is known of the changes that occur in bulk metals with aging except for calcium, which was found to shift from soft tissue with age (105). Electrolyte changes are probably related to hormonal changes that accompany the aging processes (106, 107) or to altered metal concentration per se (108, 109).

Age-dependent differences have been observed in hormonal regulation of electrolytes. For example, antidiuretic hormones enhance water excretion in young rats, but inhibit it in old ones. Hydrocortisone has similar effects, increasing water and sodium output in young rats but reducing water loss in old animals. Aldosterone increases potassium retention in old rats but essentially has no effect in very young rats (110).

Age-linked alterations in electrolyte distribution, such as loss of potassium and increased retention of sodium, was attributed to vasopressin activity (111). When vasopressin was given alone or in combination with hydrocortisone, the life span of rats was significantly increased and the mortality rate was decreased during the early months of treatment, but when it was administered to older rats, there was no effect on survival rate in spite of an improvement in electrolyte balance. Electrolyte changes with age recently have been studied by Korkusko et al. (112). Plasma, potassium, calcium, and magnesium were significantly decreased with age.

Ratios of bulk metals to ultratrace metals have been used by some workers to gain insight into the altered relationship of metal concentration in the aged (113). For example, the ratio of calcium to zinc in both trabecular and cortical bone has been found to be inversely proportional to the age of the subject. In patients with osteoporesis, minerals in general are quantitatively reduced although they remain the same qualitatively, meaning that bone density decreases as the subject advances in age.

Altered tissue concentrations of essential and nonessential ultratrace metals in man and their relationship to the aging process have been described in a series of reviews by Schroeder et al. (114, 115, 116). For example, tissue concentration of chromium decreased with age (116). Selenium did not change significantly while molybdenum was unchanged until the sixth decade of life when it tended to accumulate (115). No significant change in manganese concentration occurred in any organ with age except in the newborn (114).

Accumulation of an essential metal with age is attributed to failure of the homeostatic mechanism, increased dietary intake, or shifts in interorgan distribution mediated through changes in metabolic needs. Several examples could be cited here. Accumulation of cadmium in the aged and its link to hypertension, calcium deposition in aorta and kidney, and shifts of copper from aorta and heart to plasma and zinc accumulation in the prostate are only a few examples of how homeostasis of an ultratrace metal can be altered with age. We have examined arteriosclerotic aortas obtained from autopsy materials in a group of 18 subjects (mean age 54 years) and found them to have significantly less copper, 64 vs. 234 μg/g ash ($p < 0.001$) and zinc, 651 vs. 1196 μg/g ash ($p < 0.001$) than 22 normal aortas (mean age 57). Other metals analyzed (iron, cobalt, lead) were not significantly different (*117*). However, as with bulk metals, altered hormonal imbalance in the aged should be considered as the triggering mechanism for most of these changes, as we described earlier.

Copper and Ceruloplasmin in the Newborn and in Children. Tissue copper concentrations at birth are relatively high, particularly in the liver. However, concentration of copper in serum is lower at birth than at any other time. It is particularly high in the mother, approximately five times as much as that of the newborn (*118*). It is believed that the difference between the mother and the baby is attributable to the diminished capacity of the newborn to synthesize ceruloplasmin (*16*). Since unbound copper can pass the placental barrier, its concentration is the same on both sides of the placenta, but this is not the case with the protein-bound copper ceruloplasmin, which constitutes the majority of copper in serum.

During the first few days after birth, serum copper concentration in the baby rises from 50–150 μg/100 mL, presumably because of increases in ceruloplasmin synthesis, and then it subsides again to normal level of 100 μg/100 mL and is maintained throughout life (*119*). Thus, an infant needs at least 14 μg Cu/kg to maintain his copper balance while in children, 60–100 μg Cu/kg would be the minimal requirement. It also can be seen that low intake of copper by the mother during gestation can diminish copper stores in the embryo and thus can contribute to newborn copper deficiency.

Copper and Ceruloplasmin in the Adults. The human body contains approximately 100–150 mg Cu, most of which is located in the liver. An intake of approximately 5 mg/day is sufficient to maintain copper balance. Less than half of this copper is absorbed but most of it is excreted through the bile (*120*). Absorption from the gastrointestinal tract depends on several factors, the most important of which is probably the acidity of the intestinal content. A small portion is excreted in the urine.

Copper concentration in human tissues obtained from autopsy material varies a great deal from the time of birth to adulthood; for example, copper in brain tissue is twice as much at birth as it is at adolescence while in the liver, spleen, aorta, and lung, there is gradual decline from birth to maturity. However, copper levels in liver and aorta diminish

Table I.   Serum Copper Concentrations in Normal Male Subjects as Reported by Various Authors

First Author	Year	Method[a]	Number of Subjects	Mean ± Standard Deviation ($\mu g/100\ mL$)
Cartwright	1960	SC	120	100 ± 16
Herring	1960	SC	92	100
Rice	1962	SC	—	100 ± 12
Sunderman	1966	SC	50	121 ± 21
Sunderman	1966	AA	58	119 ± 19
Sinha	1970	AA	100	119 ± 18
Zachheim	1972	AA	33	111 ± 17
Yunice	1974	AA	180	113 ± 19

[a] SC, spectrochemical; AA, atomic absorption.

considerably after the age of 60 while in the brain, kidney, spleen, and heart, they remain unchanged (15).

Serum copper concentrations in normal subjects have been defined by several authors (95, 06, 101, 102, 121–125). Table I lists selected studies by various authors, illustrating serum copper concentrations in normal males, and Table II lists separately reported normal values in the female. Serum copper values reported by us fall within the range reported by others in both males and females. As for the difference between males and females, reports do not seem to agree. For example, Rice (126) reported higher and more reliable values in the females, while

Table II.   Serum Copper Concentrations in Normal Female Subjects as Reported by Various Authors

First Author	Year	Method[a]	Number of Subjects	Mean ± Standard Deviation ($\mu g/100\ mL$)
Cartwright	1960	SC	85	120 ± 18
Herring	1960	SC	17	117
Rice	1962	SC	—	108 ± 15
Caruthers	1966	flame	23	121 ± 11
Halsted	1968	AA	—	118 ± 21
Sinha	1970	AA	100	127 ± 21
Schenker	1971	AA	91	129
Zachheim	1972	AA	33	130 ± 23
Henkin	1973	AA	—	106 ± 18
Yunice	1974	AA	44	113 ± 20

[a] SC, spectrochemical; AA, atomic absorption.

Table III.   Serum Ceruloplasmin Concentrations in Normal Male Subjects as Reported by Various Authors

First Author	Year	Method	Number of Subjects	Mean ± Standard Deviation (mg/100 mL)
Markowitz	1955	PPD-oxidase	15	31 ± 3
Markowitz	1960	immunodiffusion	10	34 ± 4
Cox	1966	PPD-oxidase	66	30 ± 5
Haralambie	1969	immunodiffusion	18	28 ± 5
Sunderman	1970	PPD-oxidase	29	32 ± 5
Yunice	1974	immunodiffusion	180	32 ± 8

Herring et al. did not find any significant difference between the sexes (95). As can be seen from the tables, various analyses do not seem to make significant differences in the results.

Normal ranges of serum ceruloplasmin concentrations in the adult have been reported also by various investigators using a variety of analytical techniques (122, 127–132). Tables III and IV illustrate serum ceruloplasmin levels, as reported by various authors, in male and female subjects, respectively. Some authors seem to agree with our data that females tend to have a significantly higher value of ceruloplasmin than males (121, 128). In addition, Cox's data seem to indicate that serum ceruloplasmin levels gradually change from childhood through adolescence.

There is a variety of analytical methods used for ceruloplasmin determinations; the most frequently used is the p-phenylenediamine oxidase method, by virtue of its high precision. The oxidation rate of p-phenylenediamine or a derivative is measured spectrophotometrically or gasometrically, determining ceruloplasmin oxidase activity. Sunderman and Nomoto (132) determined plasma ceruloplasmin by measuring

Table IV.   Serum Ceruloplasmin Concentrations in Normal Female Subjects as Reported by Various Authors

First Author	Year	Method	Number of Subjects	Mean ± Standard Deviation (mg/100 mL)
Markowitz	1955	PPD-oxidase	15	36 ± 6
Ravin	1961	PPD-oxidase	100	32 ± 5
Caruthers	1966	PPD-oxidase	—	34
Cox	1966	—	74	31 ± 6
Laurell	1969	immunodiffusion	20	41 ± 9
Yunice	1974	immunodiffusion	44	44 ± 11

the absorption difference by ceruloplasmin at 605 mμ before and after its blue color is destroyed. Using the same technique, they obtained ceruloplasmin-bound copper indirectly by subtracting direct-reacting or unbound copper concentrations from total copper concentrations. A good correlation seems to exist between the p-phenylenediamine method and the indirect method of Sunderman and Nomoto. No such correlation between these methods and the immunodiffusion techniques has been reported in spite of the fact that the technique is commercially available and is less time consuming.

Altered Copper and Ceruloplasmin Levels as a Function of Age and Sex

In recent years, we have made a concentrated effort to study two essential ultratrace metals, copper and zinc. Since they are the most biologically active metals in the mammalian organism, their physiological role in the aging processes might be of considerable importance. Previously, we have reported from our laboratory on the role of zinc (133) and copper (134, 135) in aging. In a wide age range, 20–89 years, normal range of zinc in both serum and red blood cells was established. The data on zinc have demonstrated a linear decrease in serum zinc concentrations with increasing age and no significant changes in red blood cells. Lower levels in the females than in the males also have been shown in this study.

This chapter describes our work on serum-, copper-, and ceruloplasmin-level correlates and aging in a relatively large population. Additionally, we sought to determine if any difference exists between males and females. Since serum ceruloplasmin is closely related to copper and is known to fluctuate under various pathophysiological conditions, it also was decided to analyze the same samples for ceruloplasmin.

Materials and Methods. Subjects ranging in age between 21–89 years were randomly selected from the Domicillary Care Veterans Facility at the Central State Hospital. Determination of copper and ceruloplasmin was made on venous blood drawn from 180 males and 44 females. We have made no attempt to exclude from this study patients with chronic illnesses since it would be highly unlikely that such aged populations would be completely free of these conditions. However, patients with acute intercurrent illnesses and hepatic renal or symptomatic cardiovascular diseases were eliminated from this study. Individuals with acute illnesses as well as pregnant females or females on oral contraceptive agents also have been excluded. Some of the older patients did have evidence of atherosclerosis, mild diabetes mellitus, or compensated heart failure.

To minimize contamination from various sources, care has been taken to follow all precautions necessary, such as use of disposable plastic tubes, disposable plastic syringes, and needles. Doubly distilled water was used throughout the experiment. Serum samples were collected after centrifugation of blood and were stored in the refrigerator for copper and ceruloplasmin analysis.

Serum copper analysis was performed using a Perkin–Elmer atomic absorption spectrophotometer, Model 303. The previously described method by Prasad (*136*) was used after slight modification. For protein precipitation we used 7.5% trichloroacetic acid instead of 2N HCl used by Prasad. Analytical sensitivity for copper with this method was 0.2 μg/mL for 1% absorption and a relative detection limit of 0.005 μg/mL. Recovery studies done by adding known amounts of copper to the serum ranged between 96–104%. Analysis for ceruloplasmin was made by using commercially available immunodiffusion plates (Hyland, Inc.; normal range, 20–35 mg/100 mL). Precision of the method has been tested by running 20 determinations for both copper and ceruloplasmin on one aliquot. Coefficient of variation did not vary by more than 3–5%. All samples were run in duplicates.

Data were subjected to statistical analyses to seek differences among groups. Standard techniques were used in calculating the mean, standard deviation, correlation coefficients, and regression equations. The linear regression was calculated by the method of least squares (*137*) with various parameters for age, copper, and ceruloplasmin. The other data obtained in these experiments were analyzed by analysis of variance (Snedecor) or Duncan's Multiple Range Test (*138*) for significant differences. We have used analysis of variance to test the hypothesis that slopes of age regression lines for serum copper and ceruloplasmin and the ceruloplasmin vs. copper comparison were equal to zero for both males and females. Analysis of covariance was used to test the hypothesis that the serum copper and ceruloplasmin concentrations for males were equal to those of the females.

Results. As depicted in Table V, the mean ages, serum copper and serum ceruloplasmin concentrations ± standard deviation from 180 male and 44 female subjects were shown. As can be seen, there is no significant difference in serum copper between males and females. However, serum ceruloplasmin in the females was significantly higher than that in the males, with $p < 0.01$. This seems to agree with most of the reports mentioned earlier (Table IV). It must be pointed out that the mean age

Table V. **Mean Age, Serum Copper, and Serum Ceruloplasmin Concentrations in Male and Female Subjects**[a]

	Number of Subjects	Age (yr)	Serum Copper (μg/100 mL)	Serum Ceruloplasmin (mg/100 mL)
Male	180	51.0 ± 15.1	112.8 ± 18.9	32.1 ± 7.9
Female	44	38.7 ± 10.5	113.2 ± 20.2	43.8 ± 11.1
p	—	< 0.05	NS[b]	< 0.01

[a] ± Standard deviation.
[b] Not significant.

between male and female groups is significantly different at $p < 0.05$. From this data, it cannot be concluded whether the apparent increase in ceruloplasmin in the female is attributable to age differences or to sex.

Table VI depicts correlation coefficient, regression equations, and levels of significance for copper vs. age, ceruloplasmin vs. age, and ceruloplasmin vs. copper. As indicated by the level of significance, there was a high correlation between ceruloplasmin and copper in both males and females ($p < 0.005$). This was not surprising in view of the fact that 95% of copper in the serum is in the form of ceruloplasmin. When copper concentration was correlated with age, a significant positive correlation was obtained in the males. No such correlation was observed in the females. However, when tests of homogeneity for the regression slopes comparing males and females were applied, no significant differences were noted when comparing serum copper with age or serum ceruloplasmin with age.

Figure 1 depicts the correlations between serum copper concentrations and age in the male population. It can be seen that there is a significant increase in serum copper concentrations in males with $p < 0.025$. No such difference could be detected in the females (Figure 2). However, when corrected for age differences, the females had higher serum concentrations than males did, although the difference was not statistically significant.

Figures 3 and 4 illustrate the relationship between serum ceruloplasmin and age in both males and females, respectively. A significant correlation appears to exist in the male population whereas in the females such a correlation is lacking, but when correlation for males and females are combined, serum ceruloplasmin levels in females were significantly higher than those in males.

Table VI. Correlation Coefficient, Regression Equations, and Levels of Significance for Copper vs. Age, Ceruloplasmin vs. Age, and Ceruloplasmin vs. Copper Slopes

	Correlation Coefficient	Regression Equations[b]	p
Males (180)[a]			
Copper vs. age	0.172	$y = 101.8 + 0.215x$	< 0.025
Ceruloplasmin vs. age	0.094	$y = 29.6 + 0.049x$	NS
Ceruloplasmin vs. copper	0.355	$y = 15.5 + 0.148x$	< 0.005
Females (44)[a]			
Copper vs. age	−0.115	$y = 114.2 − 0.025x$	NS
Ceruloplasmin vs. age	−0.013	$y = 48.1 − 0.111x$	NS
Ceruloplasmin vs. copper	0.840	$y = 8.4 − 0.461x$	< 0.005

[a] Number of subjects.
[b] $y =$ 1st parameter, e.g. copper; $x =$ 2nd parameter, e.g. age.

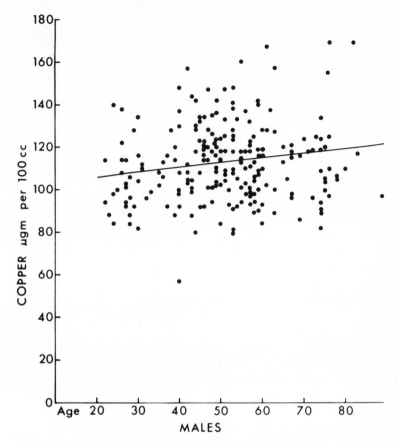

*Figure 1.   Serum copper concentrations in 180 males vs. age. The
solid line represents the calculated regression equation.*

Figure 5 depicts correlations between serum ceruloplasmin and copper concentration; as can be seen, it is significant in both males and
females with $p < 0.005$, the correlation coefficients being 0.36 and 0.84,
respectively. However, when different lots of immunodiffusion plates
were used and correlation coefficients for the samples were calculated
separately, the range of correlation in five lots was 0.30–0.88, with a mean
value of 0.53.

Discussion. Our findings of an increase in serum copper concentrations with age are in agreement with two previously reported studies,
one by Harman et al. (*139, 140*) and the other by Herring et al. (*95*).
Harman demonstrated a linear increase with age with a mean serum
concentration of 124 μg/100 mL at age 20 years and 145 μg/100 mL at
age 60. Herring reported a positive correlation in subjects at age 10–50
years with few subjects over 40 years.

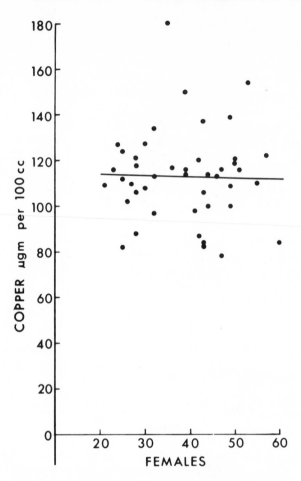

Figure 2. Serum copper concentrations in 44 females vs. age. The solid line represents the calculated regression equation.

The altered copper metabolism in the aged is not explained by our data since it did not assess copper concentrations in other body pools rich in copper, but the fact that it did increase with age in males while serum ceruloplasmin failed to do so suggests that the level of free ionic copper or copper bound to a small polypeptide molecule is higher in these subjects than the level normally found in serum. Few studies have been reported on tissue copper concentration and its relation to age except for Schroeder's work (15), alluded to earlier. He found that copper levels in liver and aorta are significantly decreased after the age of 60, whereas copper concentrations in the brain, kidney, spleen, and heart remain unchanged.

There are a variety of reasons for copper increase with age; probably, the most important is environmental exposure. As we indicated previously, the ubiquity of copper makes it unique, particularly with increased use of copper-made articles which contribute significantly to dietary, water, and air contamination. Altered homeostatis in the aged can probably be best explained by altered hormonal balance. Alterations in interorgan and intraorgan distribution depend to a large extent on cellular metabolic needs in relation to other cations, be they ultratrace or bulk metals.

Based on reports by other workers in this field and on our own findings, two hypotheses could be postulated to explain the increased serum copper concentration with age. Since tissues in these subjects were not examined for copper concentrations, increased serum copper concentration may either represent an increase in the total body pool of copper or, alternatively, the copper may have shifted from another organ, thus increasing plasma copper, plasma in this instance acting as a vehicle for transiently transporting the metal to other parts of the body where it is needed.

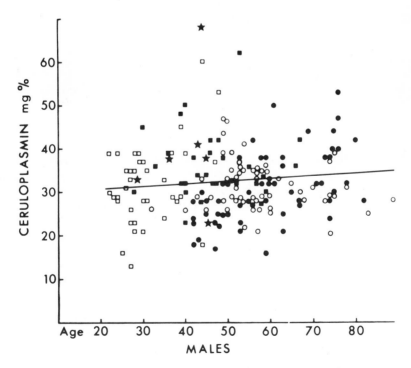

Figure 3. Serum ceruloplasmin concentrations in 180 males vs. age. The results from each of the five lots of immunodiffusion plates used are shown by different symbols.

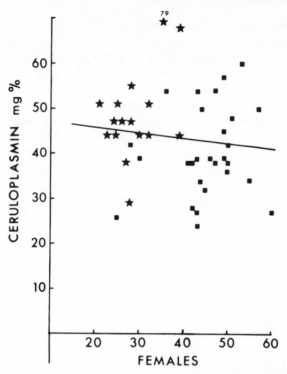

*Figure 4.   Serum ceruloplasmin concentrations in 44
females vs. age.   The results from each of the five lots
of immunodiffusion plates used are shown by different symbols.*

The first hypothesis which fits most of the criteria postulated by
Harman is the free-radical theory.   Free radicals such as hydroxy (OH)
and hydroperoxy (HO_2) are believed to be endogenously produced during the reduction of molecular oxygen.   Availability of copper and other
ultratrace metals may help to catalyze these reactions during the reduction of molecular oxygen and thus accelerate peroxidation of tissue constituents, particularly connective tissue cells, leading to bonding and
immobilization of large molecules such as DNA, enzymes, and mitochondrial lipids.   This will eventually bring about cellular dysfunction
and altered biochemical reaction in the aging processes.

It has been known previously that substances which oxidize readily,
such as ascorbic acid, are reduced in the aged (*141*) and that the use of
antioxidants such as Vitamin E and selenium compounds can inactivate
free radicals (*142*).   It also has been observed that agents which bind
ultratrace metals, particularly protein-rich sulfahydryl groups such as
cysteines and glutathione have the capability of neutralizing the effect of
free radicals (*140*).

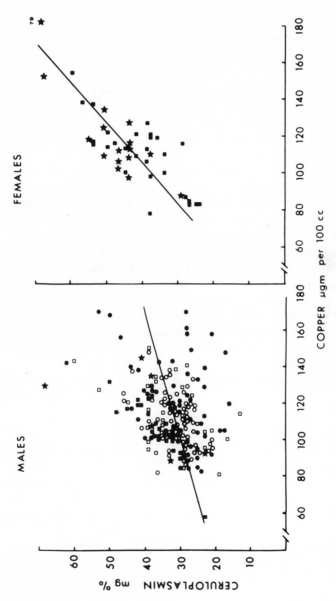

Figure 5. *Serum ceruloplasmin concentrations in 180 males and 44 females vs. copper*

It seems likely to infer that the increase in the free copper in the aged, as derived from our data, could very well make the copper available to act as the oxidation catalyst in a typical free radical reaction, as proposed by Harman.

The second hypothesis to consider is that the increase in serum copper with age might be related to changes in hormonal balance with aging. We have mentioned earlier that several authors have alluded to the role of pituitary–adrenal axis in regulating zinc and copper metabolism (97, 98, 143). A rapid and sustained drop in serum zinc following administration of large doses of corticosteroids in humans has been observed. Pregnant women and women on oral contraceptive agents show a significant increase in serum copper levels.

The chemical nature of the hormone seems to have varying effect on copper metabolism. Gonadal hormones, for example, increase ceruloplasmin by a mechanism of de novo synthesis of the protein independent of copper concentration in the liver, whereas hormones of the thyroid, adrenal, and pituitary influence copper metabolism in an indirect manner by decreasing its biliary excretion (144).

Few studies are available in the literature to correlate copper increase with age in tissues such as heart, liver, or kidneys. We recently have undertaken two studies aimed at determining altered tissue ultratrace metal concentration after hormone administration in rats. We have reported the first study (135) and made our observations, as have other workers (99), that estrogen–progesterone therapy in rats increased plasma copper concentrations and decreased zinc concentrations without significant alterations in plasma sodium, potassium, calcium, or magnesium concentrations or urinary excretion of zinc or copper. In addition, a highly significant increase of copper concentration in kidney tissue and a less significant increase in zinc have been observed. These concentrations returned to control levels two weeks after cessation of therapy. In view of the reported decreases of these hormones in the aged females (145), it is unlikely that the increased serum copper concentrations are related to the female sex hormones.

Summary and Conclusions. Serum copper and ceruloplasmin concentrations were determined in 180 male and 44 female subjects ranging in age between 20–89 years. In the male population, there was significant increase with age with $p < 0.25$. In the female population, correlation of serum copper with age was not significant. In neither the males nor the females were there any altered serum ceruloplasmin levels with age. The difference in serum copper concentration between males and females was not significant, although serum ceruloplasmin in the females was significantly higher than that in the males. The role of copper in the biological system and the environment has been reviewed in relation to other ultratrace and bulk metals, with particular reference to the aging processes. The significance and relevance of our data to current findings are interpreted in the light of alternate hypotheses as suggested by other authors. Particular emphasis was placed on hormonal imbalance and free-radical theory.

Literature Cited

1. Yunice, A. A., Lindeman, R. D., Czerwinski, A. W., Clark, M., "Effect of Age and Sex on Serum Copper and Ceruloplasmin Levels," "The Biomedical Role of Trace Elements in Aging," J. M. Hsu, Ed., pp. 55–70, Eckerd College Gerontology Center, St. Petersburg, FL, 1976.
2. Adelstein, S. J., Vallee, B. L., "Copper," *in* "Mineral Metabolism," C. L. Comar, F. Bronner, Eds., Vol. 11B, pp. 371–401, Academic, New York, 1962.
3. Carnes, W. H., "Copper and Connective Tissue Metabolism," *in* "International Review of Connective Tissue Research," D. Hall, Ed., Vol. 4, pp. 197–232, Academic, New York, 1965.
4. Cartwright, G. E., Wintrobe, M. M., "Copper Metabolism in Normal Subjects," *Am. J. Clin. Nutr.* (1964) **14**, 224–231.
5. Dowdy, R. P., "Copper Metabolism," *Am. J. Clin. Nutr.* (1969) **22**, 887–892.
6. Evans, G. W., "Copper Homeostasis in the Mammalian System," *Physiol. Rev.* (1973) **53**, 535–570.
7. Frieden, E., Hsieh, H. S., "The Biological Role of Ceruloplasmin and Its Oxidative Activity," *Adv. Exp. Med. Biol.* (1973) **74**, 505–529.
8. Frieden, E., "Ceruloplasmin, a Link Between Copper and Iron Metabolism," ADV. CHEM. SER. (1971) **100**, 292–321.
9. Graham, G. G., Cordano, A., "Copper Deficiency in Human Subjects," *in* "Trace Elements in Human Health and Disease—Zinc and Copper," A. S. Prasad, Ed., Vol. 1, pp. 363–372, Academic, New York, 1976.
10. Lee, G. R., Williams, D. M., Cartwright, G. E., "Role of Copper in Iron Metabolism and Hemebiosynthesis," *in* "Trace Elements in Human Health and Disease—Zinc and Copper," A. S. Prasad, Ed., Vol. I, pp. 373–390, Academic, New York, 1976.
11. O'Dell, B. L., "Biochemistry and Physiology of Copper in Vertebrates," *in* "Trace Elements in Human Health and Disease—Zinc and Copper," A. S. Prasad, Ed., Vol. I, pp. 381–413, Academic, New York, 1976.
12. Scheinberg, I. H., Sternlieb, I., "Copper Metabolism," *Pharm. Rev.* (1960) **12**, 355–380.
13. Scheinberg, I. H., Sternlieb, I., "Metabolism of Trace Metals," *in* "Duncan's Diseases of Metabolism, Endocrinology, and Nutrition," 6th ed., P. K. Bondy, Ed., Vol. II, pp. 1321–1334, Saunders, Philadelphia, 1969.
14. Scheinberg, I. H., "The Effects of Heredity and Environment on Copper Metabolism," *Med. Clin. North Am.* (1976) **60**, 705–712.
15. Schroeder, H. A., Nason, A. P., Tipton, I. H., Balassa, J. J., "Essential Trace Metals in Man: Copper," *J. Chronic Dis.* (1966) **19**, 1007–1034.
16. Underwood, C. J., "Trace Elements in Human and Animal Nutrition," 3rd ed., p. 22, Academic, New York, 1971.
17. Leddicotte, G. W., "Activation Analysis for Environmental Health Problems," *in* "Trace Substances in Environmental Health," D. D. Hempsill, Ed., First Annual Conference on Trace Substances in Environmental Health, pp. 56–85, University of Missouri, Columbia, 1967.

18. Tipton, I. H., Cook, M. J., Steiner, R. L., Boye, C. A., Perry, H. M., Jr., Schroeder, H. A., "Trace Elements in Human Tissue, Part 1. Methods," *Health Phys.* (1963) **9**, 89–101.
19. Salmon, M. A., Wright, T., "Chronic Copper Poisoning Presenting a Pink Disease," *Arch. Dis. Child.* (1971) **46**, 108–110.
20. Hopper, A. H., Adams, H. S., "Copper Poisoning from Vending Machines," *Public Health Rep.* (1958) **73**, 910–914.
21. McMullen, W., "Copper Contamination of Soft Drinks from Bottle Pourers," *Health Bull. (Edinburgh)* (1971) **29**, 94–96.
22. Bloomfield, J., Dixon, S. R., McCredie, S. A., "Potential Hepatotoxicity of Copper in Recurrent Hemodialysis," *Arch. Intern. Med.* (1971) **128**, 555–569.
23. Matter, B. J., Pederson, J., Psimenos, G., Lindeman, R. D., "Lethal Copper Intoxication in Hemodialysis," *Trans. Am. Soc. Artif. Intern. Organs* (1969) **15**, 309–315.
24. Pimental, J. C., Marques, F., "Vineyard Sprayers' Lung: A New Occupational Disease," *Thorax* (1969) **24**, 678–688.
25. Chantler, E., Critoph, F., Elstein, M., "Release of Copper from Copper-Bearing Intrauterine Contraceptive Devices," *Br. Med. J.* (1977) **2**, 288–291.
26. Larsson, B., Hamberger, L., "The Concentration of Copper in Human Uterine Secretion During Four Years after Insertion of a Copper-Containing Intrauterine Device," *Fertil. Steril.* (1977) **28**, 624–626.
27. Tamaya, T., Nakata, Y., Ohno, Y., Nioka, S., Furuta, N., "The Mechanism of Action of the Copper Intrauterine Device," *Fertil. Steril.* (1976) **27**, 767–772.
28. Dawson, C. R., Mallette, M. F., "The Copper Proteins," *Advan. Protein Chem.* (1945) **2**, 179–248.
29. Hart, E. B., Steenbock, H., Waddell, J., Elvehjem, C. A., "Iron in Nutrition. VII. Copper as a Supplement to Iron for Hemoglobin Building in the Rat," *J. Biol. Chem.* (1928) **77**, 797–812.
30. O'Dell, B. L., Hardwick, B. C., Reynolds, G., Savage, J. E., "Connective Tissue Defect Resulting from Copper Deficiency," *Proc. Soc. Exp. Biol. Med.* (1961) **108**, 402–405.
31. Shields, G. S., Coulson, W. F., Kimball, D. A., Carnes, W. H., Cartwright, G. E., Wintrobe, M. M., "Studies on Copper Metabolism. XXXII. Cardiovascular Lesions in Copper Deficient Swine," *Am. J. Pathol.* (1962) **41**, 603–622.
32. Marston, H. R., "Cobalt, Copper, and Molybdenum in the Nutrition of Animals and Plants," *Physiol. Rev.* (1952) **32**, 66–121.
33. Frieden, E., "The Biochemistry of Copper," *Sci. Am.* (1968) **218**, 103–114.
34. Waino, W. W., Vanderwende, C. V., Shimp, N. F., "Copper in Cytochrome C. Oxidase," *J. Biol. Chem.* (1959) **234**, 2433–2436.
35. Mann, P. J. G., "Further Purification and Properties of the Amine Oxidase of Pea Seedlings," *Biochem. J.* (1961) **79**, 623–631.
36. Buffoni, F., Blaschko, H., "Benzylamine Oxidase and Histamine Purification and Crystallization of an Enzyme from Pig Plasma," *Proc. R. Soc., Ser. B* (1964) **161**, 153–167.
37. Yamada, H., Yasunobu, K. T., "Monamine Oxidase. II. Copper, One of the Prosthetic Group of Plasma Monoamine Oxidase," *J. Biol. Chem.* (1962) **237**, 3077–3082.
38. Mondovi, B., Torillo, G., Costa, M. T., Finazzi-Agro, A., Chiancone, E., Hensen, R. E., Beinert, H., "Diamine Oxidase from Pig Kidney. Improved Purification and Properties," *J. Biol. Chem.* (1967) **242**, 1160–1167.

39. Siegel, R. C., Pinnel, S. R., Martin, G. R., "Cross-linking of Collagen and Elastin. Properties of Lysyl Oxidase," *Biochemistry* (1970) **9**, 4486–4490.
40. Craine, J. E., Daniels, G. H., Kaufman, S., "Dopamine-B-hydroxylase. The Subunit Structures and Anion Activation of the Bovine Adrenal Enzyme," *J. Biol. Chem.* (1973) **248**, 7838–7844.
41. McCord, J. M., Fridovich, I., "Superoxide Dismutase. An enzymic Function for Erythrocuprein (Hemocuprein)," *J. Biol. Chem.* (1969) **244**, 6049–6055.
42. Markowitz, H., Cartwright, G. E., Wintrobe, M. M., "Studies of Copper Metabolism. XXVII. Isolation and Properties of Erythrocyte Cuproprotein (Erythrocuprein)," *J. Biol. Chem.* (1959) **234**, 40–45.
43. Porter, H., Ainsworth, S., "The Isolation of the Copper-Containing Protein Cerebrocuprein I from Normal Human Brain," *J. Neurochem.* (1959) **5**, 91–98.
44. Porter, H., Sweeney, M., Porter, E. M., "Human Hepatocuprein. Isolation of a Copper Protein from the Subcellular Soluble Fraction in Adult Human Liver," *Arch. Biochem. Biophys.* (1964) **16**, 319–325.
45. Keele, B. B., McCord, J. M., Fridovich, I., "Further Characterization of Bovine Superoxide Dismutase and Its Association from Bovine Heart," *J. Biol. Chem.* (1971) **246**, 2875–2880.
46. Weiseger, R. A., Fridovich, I., "Superoxide Dismutase. Organelle Specificity," *J. Biol. Chem.* (1973) **248**, 3582–3592.
47. Holmberg, C. G., Laurell, C .B., "Investigations in Serum Copper. II. Isolation of the Copper-Containing Protein and a Description of Some of Its Properties," *Acta Chem. Scand.* (1948) **2**, 550.
48. Gubler, C. J., Lahey, M. E., Cartwright, G. E., Wintrobe, M. M., "Studies on Copper Metabolism. IX. The Transportation of Copper in Blood," *J. Clin. Invest.* (1953) **32**, 405–414.
49. Neumann, P. Z., Sass-Kortsak, I., "The State of Copper in Human Serum: Evidence for an Amino Acid-Bound Fraction," *J. Clin. Invest.* (1967) **46**, 646–658.
50. Lahey, M. E., Gubler, C. J., Chase, M. S., Cartwright, G. E., Wintrobe, M. M., "Studies of Copper Metabolism. II. Hematologic Manifestations of Copper Deficiency in Swine," *Blood* (1952) **7**, 1053–1074.
51. Gubler, C. J., "Absorption and Metabolism of Iron," *Science* (1956) **123**, 87–90.
52. Curzon, G., O'Reilly, S., "A Coupled Iron Ceruloplasmin Oxidation System," *Biochem. Biophys. Res. Commun.* (1960) **2**, 284–286.
53. Osaki, S., "Kinetic Studies of Ferrous Iron Oxidation with Crystalline Human Ferroxidase (Ceruloplasmin)," *J. Biol. Chem.* (1966) **241**, 5053–5059.
54. Frieden, E., "The Ferrous to Ferric Cycles in Iron Metabolism," *Nutr. Rev.* (1973) **31**, 41–44.
55. Gaber, B. P., Aisen, P., "Is Bivalent Iron Bound to Transferrin?" *Biochim. Biophys. Acta* (1970) **221**, 228–233.
56. Roeser, H. P., Lee, G. R., Nacht, S., Cartwright, G. E., "The Role of Ceruloplasmin in Iron Metabolism," *J. Clin. Invest.* (1970) **49**, 2408–2417.
57. Williams, D. M., Loukopoulos, D., Lee, G. R., Cartwright, G. E., "Role of Copper in Mitochondrial Iron Metabolism," *Blood* (1976) **48**, 77–85.
58. O'Dell, B. L., Birk, D. W., Ruggles, D. L., Savage, J. E., "Composition of Aortic Tissue from Copper Deficient Chicks," *J. Nutr.* (1966) **88**, 9–14.
59. Coulson, W. F., Carnes, W. H., "Cardiovascular Studies on Copper-Deficient Swine. V. The Histogenesis of the Coronary Artery Lesions," *Am. J. Pathol.* (1963) **43**, 945–954.

60. Rucker, R. B., Murray, J., Riggins, R. S., "Nutritional Copper Deficiency and Penicillamine Administration. Some Effects on Bone Collagen and Arterial Elastin Cross-Linking," *Adv. Exp. Med. Biol.* (1977) **86B**, 619–648.

61. Everson, G. J., Schrader, R. E., Wang, T. J., "Chemical and Morphological Changes in the Brains of Copper-Deficient Guinea Pigs," *J. Nutr.* (1968) **96**, 115–118.

62. Danks, D. M., Campbell, P. E., Walker-Smith, J., Stevens, J. B., Gillespie, J. M., Bloomfield, J., Turner, B., "Menkes-Kinky Hair Syndrome," *Lancet* (1972) **1**, 1000–1103.

63. Wintrobe, M. M., "Clinical Hematology," 5th ed., p. 141, Lea and Febiger, Philadelphia, 1961.

64. Holtzman, N. A., Charache, A., Cordano, A., Graham, G. G., "Distribution of Serum Copper in Copper Deficiency," *John Hopkins Med. J.* (1970) **126**, 34–42.

65. Cordano, A., Baertl, J. M., Graham, G. G., "Copper Deficiency in Infancy," *Pediatrics* (1964) **34**, 324–336.

66. Dauncey, M. J., Shaw, J. C. L., Urman, J., "The Absorption and Retention of Magnesium, Zinc, and Copper by Low Birth Weight Infants Fed Pasteurized Human Breast Milk," *Pediatr. Res.* (1977) **11**, 1033–1039.

67. Al-Rashid, R. A., Spangler, J., "Neonatal Copper Deficiency," *N. Engl. J. Med.* (1971) **285**, 841–843.

68. Karpel, J. T., Peden, V. H., "Copper Deficiency in Long-term Parenteral Nutrition," *J. Pediatr.* (1972) **80**, 32–36.

69. Sivagubramanian, K. N., Hoy, G., Davitt, M. K., Henkin, R. I., "Zinc and Copper Changes after Neonatal Parenteral Alimentation," *Lancet* (1978) **1**, 508.

70. Dunlap, W. M., James, G. W., Hume, D. M., "Anemia and Neutropenia Caused by Copper Deficiency," *Ann. Intern. Med.* (1974) **80**, 470–476.

71. Solomons, N. W., Layden, T. J., Rosenberg, I. H., Vo-Khactu, K., Sandstead, H. H., "Plasma Trace Metals During Total Parenteral Alimentation," *Gastroenterology* (1976) **70**, 1022–1025.

72. Fleming, C. R., Hodges, R. F., Hurley, L. S., "A Prospective Study of Serum Copper and Zinc Levels in Patients Receiving Total Parenteral Nutrition," *Am. J. Clin. Nutr.* (1976) **29**, 70–77.

73. Williams, D. M., Atkins, C. L., Frens, D. B., Bray, P. F., "Menkes-Kinky Hair Syndrome. Studies of Copper Metabolism and Long Term Copper Therapy," *Pediatr. Res.* (1977) **11**, 823–826.

74. Bearn, A. G., "Wilson's Disease. An Inborn Error of Metabolism with Multiple Manifestations," *Am. J. Med.* (1957) **22**, 747–757.

75. Walshe, J. M., "Wilson's Disease: A Review. Biochemistry of Copper," J. Peisach, et al., Eds., pp. 475–498, Academic, New York, 1966.

76. Uzman, L. L., Iber, F. L., Chalmers, T. C., Knowlton, M., "The Mechanism of Copper Deposition in the Liver in Hepatolenticular Degeneration—Wilson's Disease," *Am. J. Med. Sci.* (1956) **231**, 511–518.

77. Scheinberg, I. H., Gitlin, D., "Deficiency of Ceruloplasmin in Patients with Hepatolenticular Degeneration (Wilson's Disease)," *Science* (1952) **116**, 484–485.

78. Sternlieb, I., "Evolution of the Hepatic Lesion in Wilson's Disease (Hepatolenticular Degeneration)," *Prog. Liver Dis.* (1972) **4**, 511–525.

79. Goldfischer, S., Sternlieb, I., "Changes in the Distribution of Hepatic Copper in Relation to the Progression of Wilson's Disease (Hepatolenticular Degeneration)," *Am. J. Pathol.* (1968) **53**, 883–901.

80. Osborn, S. B., Walshe, J. M., "Studies with Radioactive Copper (Cu^{64} and Cu^{67}) in Relation to the Natural History of Wilson's Disease," *Lancet* (1967) **1**, 346–350.

81. Frommer, D. J., "The Measurement of Biliary Copper Secretion in Humans," *Clin. Sci.* (1972) **42**, 26P.
82. Scheinberg, I. H., Sternlieb, I., "Wilson's Disease," *Annu. Rev. Med.* (1965) **16**, 119–134.
83. Fleming, C. R., Dickson, E. R., Wahner, H. W., Hollenhorst, R. W., McCall, J. T., "Pigmented Corneal Rings in Non-Wilsonian Disease," *Ann. Intern. Med.* (1977) **86**, 285–288.
84. LaRusso, N. F., Summerskill, W. H., McCall, J. T., "Abnormalities of Chemical Types for Copper Metabolism in Chronic Active Liver Disease: Differentiation from Wilson's Disease," *Gastroenterology* (1976) **70**, 653–655.
85. Ritland, S., Steinnes, E., Skredes, S., "Hepatic Copper Content, Urinary Copper Excretion and Serum Ceruloplasmin in Liver Disease," *Scand. J. Gastroenterol.* (1977) **12**, 81–88.
86. Scott, J., Gollan, J. L., Soumarian, S., Sherlock, S., "Wilson's Disease Presenting as Chronic Active Hepatitis," *Gastroenterology* (1978) **74**, 645–651.
87. Vierling, J. M., Shrager, R., Rumble, W. F., Aamodt, R., Berman, M. D., Jones, E. A., "Incorporation of Radiocopper into Ceruloplasmin in Normal Subjects and in Patients with Primary Biliary Cirrhosis and Wilson's Disease," *Gastroenterology* (1978) **74**, 652–660.
88. Walshe, J. M., Potter, G., "The Pattern of the Whole Body Distribution of Radioactive Copper (Cu^{67} and Cu^{64}) in Wilson's Disease and Various Control Groups," *Q. J. Med.* (1977) **46**, 445–462.
89. Bearn, A. G., "Genetic and Biochemical Aspects of Wilson's Disease," *Am. J. Med.* (1953) **15**, 442–449.
90. Bearn, A. G., "Wilson's Disease," *in* "The Metabolic Basis of Inherited Disease," J. B. Stanbury, J. C. Syngaarden, D. S. Fredrickson, Eds., 3rd ed., pp. 1033–1050, McGraw-Hill, New York, 1972.
91. McCullars, G. M., O'Reilley, S., Brennan, M., "Pigment Binding of Copper in Human Bile," *Clin. Chim. Acta.* (1977) **74**, 33–38.
92. Evans, G. W., Cornatzer, W. E., Dubois, R. S., Hambidge, K. M., "Characterization of Hepatic Copper Proteins from Mammalian Species and a Human with Wilson's Disease," *Fed. Proc.* (1971) **30**, 461.
93. Ryden, L., Deutsch, H. F., "Preparation and Properties of the Major Binding Component in Human Fetal Liver. Its Identification as Metallothionein," *J. Biol. Chem.* (1978) **253**, 519–524.
94. Owen, C. A., Jr., Hazelrig, J. B., "Copper Deficiency and Copper Toxicity in the Rat," *Am. J. Physiol.* (1968) **215**, 334–338.
95. Herring, B. W., Leavell, B. S., Paixao, L. M., Yoe, J. H., "Trace Metals in Human Plasma and Red Blood Cells. I. Observations of Normal Subjects," *Am. J. Clin. Nutr.* (1960) **8**, 846–854.
96. Sinha, S. N., Gabrielli, E. R., "Serum Copper and Zinc Levels in Various Pathological Conditions," *Am. J. Clin. Pathol.* (1970) **54**, 570–577.
97. Flynn, A., Pories, W. J., Strain, W. H., Hill, O. A., Jr., "Mineral Elements Correlation with Adenohypophyseal-Adrenal Cortex Function and Stress," *Science* (1971) **173**, 1035–1036.
98. Briggs, M. H., Briggs, M., Austin, J., "Effects of Steroid Pharmaceuticals on Plasma Zinc," *Nature* (1971) **232**, 480.
99. Sato, N., Henkin, R. I., "Pituitary-Gonadal Regulation of Copper and Zinc Metabolism in the Female Rats," *Am. J. Physiol.* (1973) **225**, 508–512.
100. Briggs, M., Austin, J., Staniford, M., "Oral Contraceptives and Copper Metabolism," *Nature* (1970) **225**, 81.
101. Schenker, J. G., Hellerstein, S., Jungries, E., Polishuk, W. Z., "Serum Copper and Zinc Levels in Patients Taking Oral Contraceptives," *Fertil. Steril.* (1971) **22**, 229–234.

102. Halsted, J. A., Hackley, B. M., Smith, J. C., Jr., "Plasma, Zinc, and Copper in Pregnancy and After Oral Contraceptives," *Lancet* (1968) **2**, 278–279.

103. Evans, G. W., Wiederanders, R. D., "Pituitary-Adrenal Regulation of Ceruloplasmin," *Nature* (1967) **215**, 766–767.

104. Schroeder, H. A., "The Biological Trace Elements or Peripatetics through the Periodic Table," *J. Chronic Dis.* (1965) **18**, 217–228.

105. Lansing, A. I., "Increase of Cortical Calcium with Age in the Cells of a Rotifer, *Cuchlanis dilatata*, a Planerian, *Phacogata Sp.*, and a Toad, *Bufo fowleri* as Shown by the Microincineration Technique," *Biol. Bull.* (1942) **82**, 392–400.

106. Friedman, S. M., Hinke, J. A. M., Friedman, C. L., "Neurohypophyseal Responsiveness in the Normal and Senescent Rats," *J. Gerontol.* (1956) **11**, 286–291.

107. Friedman, S. M., Nakashima, N., Friedman, C. L., "Prolongation of Lifespan in the Old Rat by Adrenal and Neurohypophyseal Hormones," *Gerontologia* (1965) **11**, 129–140.

108. Dunihue, F. W., "Reduced Juxtaglomerular Cell Granularity, Pituitary Neurosecretory Material and Width of the Zona Glomerulosa in Aging Rats," *Endocrinology* (1965) **77**, 948–951.

109. Rodeck, H. K., Lederis, K., Heller, H., "The Hypothalamo-Neurohypophyseal System in Old Rats," *J. Endocrinol.* (1960) **21**, 225–228.

110. Bellamy, D., "Hormonal Effects in Relation to Aging in Mammals," *Symposium for Experimental Biology* (1967) **21**, 427–453.

111. Friedman, S. M., Friedman, C. L., "Effects of Posterior Pituitary Extracts on the Life Span," *Nature* (1963) **200**, 237–138.

112. Korkusko, Von O. V., Kupras, L. P., Kusenov, R. K., "Altersbesonderheiten des Wasser-Elektrolyt-Austausches," *Z. Alternsforsch.* (1977) **32**, 181–186.

113. Aitkin, J. M., "Factors Affecting the Distribution of Zinc in the Human Skeleton," *Calcif. Tissue Res.* (1976) **20**, 23–30.

114. Schroeder, H. A., Balassa, J. J., Tipton, I. H., "Essential Trace Metals in Man: Manganese, a Study in Homeostasis," *J. Chronic Dis.* (1966) **19**, 545–571.

115. Schroeder, H. A., Balassa, J. J., Tipton, I. H., "Essential Trace Metals in Man: Molybdenum," *J. Chronic. Dis.* (1970) **23**, 481–499.

116. Schroeder, H. A., Balassa, J. J., Tipton, I. H., "Abnormal Trace Metals in Man: Chromium," *J. Chronic. Dis.* (1962) **15**, 941–964.

117. Yunice, A. A., Keim, T., Kirk, J. E., Perry, H. M., Jr., "Decreased Zinc and Copper Content of Arterioloscierotic Aortas," *Clin. Res.* (1969) **17**, 65.

118. Scheinberg, I. H., Cook, C. D., Murphy, J. A., "The Concentration of Copper and Ceruloplasmin in Maternal and Infant Plasma at Delivery," *J. Clin. Invest.* (1954) **33**, 963.

119. Widdowson, E. M., "Trace Elements in Human Development," *in* "Mineral Metabolism in Pediatrics," D. Baltrop, W. L. Burland, Eds., pp. 85–98, F. A. Davis, Philadelphia, 1969.

120. Bush, J. A., Mahoney, J. P., Markowitz, H., Gubler, C. J., Cartwright, G. E., Wintrobe, M. M., "Studies on Copper Metabolism. XVI. Radioactive Copper Studies in Normal Subjects and in Patients with Hepatolenticular Degeneration," *J. Clin. Invest.* (1955) **34**, 1766–1778.

121. Cartwright, G. E., Markowitz, H., Shields, G. S., Wintrobe, M. M., "Studies in Copper Metabolism. XXIX. A Critical Analysis of Serum Copper and Ceruloplasmin Concentrations in Normal Subjects, Patients with Wilson's Disease," *Am. J. Med.* (1960) **28**, 555–563.

122. Caruthers, M. E., Hobbs, C. B., Warren, R. L., "Raised Serum Copper and Ceruloplasmin Levels in Subjects Taking Oral Contraceptives," *J. Clin. Pathol.* (1966) **19**, 498–500.
123. Sunderman, F. W., Jr., Roszel, N. O., "Measurements of Copper in Biologic Materials by Atomic Absorption Spectrophotometry," *Am. J. Clin. Pathol.* (1967) **48**, 286–294.
124. Zachheim, H. S., Wolf, P., "Serum Copper in Psoriasis and Other Dermatoses," *J. Invest. Dermatol.* (1972) **58**, 28–32.
125. Henkin, R. I., Schulman, J. D., Schulman, D. B., Bronzert, D. A., "Changes in Total, Non-Diffusible, and Diffusible Plasma Zinc and Copper During Infancy," *J. Pediatr.* (1973) **82**, 831–837.
126. Rice, E. W., "Spontaneous Variations in the Concentration of Serum Copper of Healthy Adults," *Am. J. Med. Sci.* (1962) **102**, 593–597.
127. Ravin, H. A., "An Improved Colorimetric Enzymatic Assay of Ceruloplasmin," *J. Lab. Clin. Med.* (1961) **58**, 161–168.
128. Cox, D. W., "Factors Influencing Serum Ceruloplasmin Levels in Normal Individuals," *J. Lab. Clin. Med.* (1966) **68**, 893–904.
129. Haralambie, G., "Zum Normalevert des Coeruloplasmins in Serum and Seine Beeinflussing durch Korperarbeit," *Z. Klin. Chem. Klin. Biochem.* (1969) **7**, 352–355.
130. Laurell, C. B., Kullander, S., Thorell, J., "Plasma Proteins after Continuous Oral Use of a Progesterone as a Contraceptive," *Scand. J. Clin. Lab. Invest.* (1969) **24**, 387–389.
131. Morell, A. G., Windsor, J., Steinlieb, I., Scheinberg, I. H., "Measurement of the Concentration of Ceruloplasmin in Serum by Determination of Its Oxidase Activity," in "Laboratory Diagnosis of Liver Disease," F. W. Sunderman, F. W. Sunderman, Jr., Eds., 193–195, W. H. Green, St. Louis, MO, 1968.
132. Sunderman, F. W., Jr., Nomoto, S., "Measurement of Human Serum Ceruloplasmin by *p*-Phenylenediamine Oxidase Activity," *Clin. Chem. Winston-Salem, N.C.* (1970) **16**, 903–910.
133. Lindeman, R. D., Clark, M. D., Colmore, J. P., "Influence of Age and Sex on Plasma and Red Cell Zinc Concentrations," *J. Gerontol.* (1971) **26**, 358–363.
134. Yunice, A. A., Lindeman, R. D., Czerwinski, A. W., Clark, M., "Influence of Age and Sex on Serum Copper and Ceruloplasmin Levels," *J. Gerontol.* (1974) **29**, 277–281.
135. Yunice, A. A., Lindeman, R. D., "Effect of Estrogen–Progesterone Administration on Tissue Cation Concentrations in the Rat," *Endocrinology* (1975) **97**, 1263–1269.
136. Prasad, A. S., Oberleas, D., Halsted, A. J., "Determination of Zinc in Biological Fluids by Atomic Absorption Spectrophotometery in Normal and Cirrhotic Subjects," *J. Lab. Clin. Med.* (1965) **66**, 508–516.
137. Snedecor, G. W., Cochran, W. G., "Statistical Methods," 6th ed., pp. 91–119, 135–171, Iowa State University, Ames, 1967.
138. Duncan, D. B., "Multiple Range and Multiple F Tests," *Biometrics* (1955) **11**, 1–42.
139. Harman, D., "The Free Radical Theory of Aging: Effect of Age on Serum Copper Levels," *J. Gerontol.* (1965) **20**, 151–153.
140. Harman, D., "Free Radical Theory of Aging: Effect of Free Radical Reaction Inhibitors on the Mortality Rate of Male LAF Mice," *J. Gerontol.* (1968) **23**, 476–482.
141. Kirk, J. E., "Blood and Urine Vitamin Levels in the Aged," Nat. Vitamin Foundation, Inc., Symposium on Problems of Gerontology, 1954, New York, pp. 73–94.
142. Tappel, A. L., "Will Antioxidant Nutrients Slow Aging Processes?" *Geriatrics* (1968) **23**, 97–105.

143. Henkin, R. I., Meret, S., Jacobs, J. B., "Steroid-dependent Changes in Copper and Zinc Metabolism," *J. Clin. Invest.* (1969) **48**, 38a.
144. Evans, G. W., Cornatzer, N. F., Cornatzer, W. E., "Mechanism for Hormone Induced Alterations in Serum Ceruloplasmin," *Am. J. Physiol.* (1970) **218**, 613–615.
145. Pincus, G., Dorfman, R. I., Ramanoff, L. P., Rubin, B. L., Block, E., Calvo, J., Freeman, H., "Recent Progress in Hormone Research," Vol. XI, Gregory Pincus, Ed., 307–334, Academic, New York, 1955.

RECEIVED May 12, 1978. This study was supported by the Medical Research Service of the Veteran's Administration Hospital.

INDEX

The text of this book is set in 10 point Caledonia with two points of leading. The chapter numerals are set in 30 Point Garamond; the chapter titles are set in 18 point Garamond Bold.

The book is printed offset on Text White Opaque 50-pound. The cover is Joanna Book Binding blue linen.

Jacket design by Alan Kahan. Editing and production by Saundra Goss.

The book was composed by Service Composition Co., Baltimore, MD., printed and bound by The Maple Press Co., York, PA.